The Modernization of the Louisiana Sugar Industry 1830–1910

The Modernization of the Louisiana Sugar Industry 1830–1910

John Alfred Heitmann

Louisiana State University Press
Baton Rouge and London

Designer: Christopher Wilcox
Typeface: Trump Mediaeval
Typesetter: G & S Typesetters, Inc.
Printer: Thomson-Shore, Inc.
Binder: John Dekker and Sons, Inc.

10 9 8 7 6 5 4 3 2 1

Library of Congress Cataloging-in-Publication Data
Heitmann, John Alfred.
    The modernization of the Louisiana sugar industry, 1830–1910.

    Bibliography: p.
    Includes index.
    1. Sugar trade—Louisiana—Technological innovations—History.
I. Title.
HD9107.L8H45   1987   338.4'56641'09763   86-27639
ISBN 0-8071-1324-7

For Kaye

Contents

Illustrations

Acknowledgments

I would like to thank Professor Owen Hannaway of Johns Hopkins University for his encouragement and guidance during the preparation of this manuscript. From Professor Hannaway I learned not only the fundamentals of the history of chemistry but also how to think as a historian. He gave of his time freely and challenged me to reanalyze my source materials and to refine my ideas.

Professors Robert Kargon, Stewart W. Leslie, and Louis Galambos were also instrumental in shaping my investigation of the Louisiana sugar industry. Dr. Kargon inspired me to pursue research in several areas that proved fruitful, and he instilled an excitement in studying the past that has remained with me. Professor Leslie constantly reminded me that incisive ideas are the basis of good history. Just as important, however, his friendship helped sustain me during the trials of revision. Dr. Galambos instructed me in the basics of economic history; his organizational approach has had a lasting effect on my work. In addition to my mentors, Thomas Cornell, a fellow graduate student at Johns Hopkins and now at the Rochester Institute of Technology, read preliminary drafts of chapters and frequently made suggestions that helped broaden my perspective.

Several institutions either sponsored or cooperated with my research and writing efforts. The Johns Hopkins University provided me with financial assistance during my graduate studies. The University of Dayton provided me with a generous summer fellowship to complete the revisions of this manuscript and awarded me a travel grant that enabled me to visit repositories in Louisiana and Washington, D.C. Also, many of my colleagues in the history department at the University of Dayton, including Roberta Alexander, Pat Palermo, and Frank Mathias, encouraged me to pursue my work during a difficult first year of full-time teaching.

I am also grateful for the assistance of archivists at the Manuscript Division of the Library of Congress, the National Archives, the College of William and Mary, the Southern Historical Collection at the University of North Carolina at Chapel Hill, Tulane University, the University of Southwestern Louisiana, and the Historic New Orleans Collection. In particular, Stone Miller, Judy Bolton, and Gisela Losada at Special Collections, Hill Memorial Library, Louisiana State University, Steve Reinhart at the Louisiana State Museum, and Alfred Lemmon of the Historic New Orleans Collection were most helpful in locating manuscripts and photographs. In addition, this work would not have been written without the use of the extensive collections on sugar chemistry and technology housed at the National Agricultural Library, where Dr. Alan Fusonie not only helped me locate materials, but also made me feel welcome during my many hours of research.

Finally, I thank my family for their love and support. My mother, Caroline, typed many of the chapters of the final manuscript. And Kaye, my wife, who pushed me to continue when I became discouraged, served as my editor. For her contributions and reassurance, this work is dedicated to Kaye.

The Modernization of the Louisiana Sugar Industry 1830–1910

Preface

Gerald D. Nash, writing in the *Journal of Southern History* more than twenty years ago, lamented the fact that scholars of his generation were excessively preoccupied with traditional themes in southern history, namely politics and race relations.[1] In his opinion, historians were ignoring the invisible yet powerful economic forces that contributed so much to shaping the development of the South. Nash called for a new approach in 1966, one focusing on the process of industrialization in the South after 1880 and the associated internal dynamics that eventually resulted in revolutionary economic and societal changes. He suggested several topics for study, including financial institutions, trade associations, the role of state and federal government in the economy, entrepreneurial activities, and above all the influence of industry in shaping modern southern culture.

While Nash's article was well received and is frequently cited, only a handful of historians have heeded his advice. With the notable exception of James C. Cobb's *Industrialization and Southern Society, 1877–1984* and a few other publications, political history continues to dominate the field.[2] Traditionally, Louisiana historians have been

1. Gerald D. Nash, "Research Opportunities in the History of the South After 1880," *Journal of Southern History,* XXXII (1966), 308–24.

2. James C. Cobb, *Industrialization and Southern Society, 1877–1984* (Lexington, Ky., 1984); Raymond Arsenault, "The End of the Long Hot Summer: The Air Con-

preoccupied with the political caldron in that state, and works such as William Ivey Hair's *Bourbonism and Agrarian Protest: Louisiana Politics, 1877–1900*, Alan P. Sindler's *Huey Long's Louisiana: State Politics, 1920–1952*, Perry H. Howard's *Political Tendencies in Louisiana*, and T. Harry Williams' *Huey Long* all reflect this tendency.[3] Undoubtedly, the political machinations taking place in Louisiana, which sometimes can be compared to the goings on in a banana republic, provide the basis for instructive and entertaining stories. However, other tales need to be told as well. Entire areas of scholarship remain neglected. For example, there exists no comparable history of the development of Louisiana's chemical industry, certainly as significant as politics in shaping the state's character. Just as Nash warned two decades ago, our historical understanding of the South, and in particular Louisiana, continues to be distorted. Old ground has been repeatedly retilled, while vast new areas remain to be explored.

This study of an industry in transition—the late-nineteenth-century Louisiana sugar industry—departs from the conventional in several ways. I shall take up Nash's challenge and extend it by examining the process of industrialization and by investigating its scientific and technological basis. My study is built upon the work of many able scholars, and in particular that of J. Carlyle Sitterson.[4] Sitterson's comprehensive *Sugar Country* taught me a great deal about the Louisiana sugar industry; however, it also prompted questions that eventually led me on a different path from that of its author. Although Sitterson touched upon science and technology and suggested that institutions were important to the modernization of the sugar industry, he did not fully characterize their significance.[5]

---

ditioner and Southern Culture," *Journal of Southern History*, L (1984), 597–628; David O. Whitten, *The Emergence of Giant Enterprise, 1860–1914* (Westport, Conn., 1983).

3. For an excellent review of the historiography related to Louisiana, see Light Townsend Cummins and Glen Jeansonne (eds.), *A Guide to the History of Louisiana* (Westport, Conn., 1982).

4. J. Carlyle Sitterson, *Sugar Country: The Cane Sugar Industry in the South, 1753–1950* (Lexington, Ky., 1953).

5. See J. Carlyle Sitterson, "Expansion, Reversion, and Revolution in the Southern Sugar Industry, 1850–1910," *Bulletin of the Business Historical Society*, XXVII (1953), 137.

When these themes are carefully studied, a deeper understanding of the process of change and a new view of the Louisiana sugar bowl emerge.

In this study of science, technology, and the modernization of the Louisiana sugar industry, the key unit of analysis is the organization.[6] Institutions are created in response to specific conditions existing within a culture, and their formation is reflective of dynamic processes taking place within society. Indeed, trade organizations, experiment stations, and university programs related to the sugar industry in nineteenth-century Louisiana were a product of a complex environment; they also concurrently sustained economic, social, and political change. In general, these institutions were established for the purpose of achieving stability and control over an industry threatened by foreign competition and confronted with the problems of a new labor system.

An organizational approach to this period allows a dynamic Louisiana sugar industry to emerge within a broad context. Rather than placing developments in the Louisiana sugar bowl in a relatively isolated southern tradition, as in *Sugar Country*, events are interpreted within an international perspective. Local institutions served as transmitters and facilitators of knowledge and expertise originating in Europe, and they catalyzed the process of assimilation that contributed so much to industrial modernization in Louisiana. Prior to the Civil War, France proved to be the source of many new technical methods and scientific ideas. After 1865 the European beet-sugar industry, and particularly German processes and chemical controls, contributed to the dramatic changes in Louisiana.

While institutions serve as the gatekeepers of knowledge and are thus the focal point of this study, the story of science, technology, and the development of the Louisiana sugar industry is far from bloodless. The effectiveness of these organizations was the consequence of individual initiative. Between 1877 and 1900 prominent planters, merchants, and manufacturers worked hard to create and sustain local institutions and to promote the idea that application of scientific knowledge resulted in economic profits. In a 1982 ar-

6. Louis Galambos, "Technology, Political Economy, and Professionalization: Central Themes of the Organizational Synthesis," *Business History Review*, LVII (1983), 471–93.

ticle, Louis Ferleger focused on the tradition-bound planter in his interpretation of the innovations in sugar technology; on the contrary, the elite planters involved in manufacturing were in reality more cosmopolitan than parochial and more progressive than conservative.[7]

My interest in the Louisiana sugar industry had its origins in previous research dealing with the history of chemical engineering. As I reviewed the literature for the first American exposition concerning the discipline, I found a 1900 address by Magnus Swenson to faculty and students at the University of Wisconsin that outlined the role and value of the chemical engineer. Swenson, a former Wisconsin professor and university trustee, drew on his experiences in the Louisiana sugar industry and pointed to developments there as being crucial to the emergence of the theoretical and practical basis of this applied science. Swenson noted that chemical control of the sugarhouse was rarely practiced until rapidly falling prices during the 1880s "made it a question of life and death to the industry." Because the sugar chemists were limited in their training, large amounts of capital were wasted on inadequate machinery. Within a few years, however, large-scale yields of sugar were improved "beyond what was thought possible." Swenson attributed *"this . . . to the gradual evolution of the chemical engineer who has both the necessary chemical knowledge and the technical training."* He credited these improvements in the sugar trade "in a very large measure to the Audubon Sugar School."[8] Swenson's statements were what I was looking for. It seemed to me that by examining the work of applied scientists developing process modifications, as well as the Audubon Sugar School curriculum and the Louisiana Sugar Experiment Station research programs, the emergence of a new profession within an industrial context could be traced.

Beginning with a search for the origins of a hybrid body of knowledge and a profession, I was thus led to a unique educational institution for its time, the Audubon Sugar School, which was established in 1891 in New Orleans. What was so intriguing to me was that

7. Louis Ferleger, "Farm Mechanization in the Southern Sugar Sector After the Civil War," *Louisiana History,* XXIII (1982), 21–34.

8. Magnus Swenson, "The Chemical Engineer," *Bulletin of the University of Wisconsin, Engineering Series,* II (1900), 199–200.

chemical engineering, a discipline closely associated with the modern chemical industry, had one wing of its development in a most unlikely setting, the Deep South, and was related to an industry normally characterized as conservative and resistant to change. What were the circumstances surrounding the establishment of the Audubon Sugar School? This and related questions were answered by studying the Louisiana Sugar Planters' Association (LSPA), the crucial organization involved in the modernization of the state's sugar industry.

Established in 1877, the LSPA's immediate goal was to secure favorable tariff legislation by acting as a lobby for the industry in Washington. Once this mission was accomplished, however, the LSPA leadership began to recognize the potential benefits of the application of chemical and engineering knowledge to sugar manufacture. They then concentrated their efforts on incorporating new technological innovations and chemical methods with the purpose of increasing production efficiency. However, by 1880 the LSPA membership realized that their lack of scientific expertise could prevent them from reaching this objective. To overcome the deficiency, the association pursued several strategies to secure university-trained scientists and engineers for the Louisiana sugar industry. In obtaining experts, the LSPA embarked upon a program of institution building and formed a network of organizational structures that in part sustained the local economy.[9] A scientific and technical institutional infrastructure emerged by 1890 that possessed an inherent flexibility to transcend from one chemical technology to another, from sugar to petroleum, as Louisiana moved from the nineteenth into the twentieth century.

Standing on the balcony of Duncan Kenner's nineteenth-century Ascension Parish plantation home, one can sense the distinctions between the past and the present. In the foreground under the shade of huge, ancient oak trees stand primitive grinding equipment and large open kettles that were once used to boil cane syrup. Yet, on the terrain beyond the plantation loom smokestacks, tanks, and frac-

9. On systems and networks and their application to historical analysis, see Thomas P. Hughes, *Networks of Power: Electrification in Western Society* (Baltimore, 1983), 5–6.

tionating columns behind the gates of a Geigy Corporation petro-
chemical plant. While travel guides rarely mention the presence of
chemical plants, it is the modern chemical industry that dominates
the region's landscape, that has paved a road of prosperity for its citizens, and that most likely will determine its future economy. And as
this study will show, the efforts of scientists and engineers and
prominent local businessmen, the active role of the federal government, and the interplay of local institutions are the keys to sustained economic growth in a dynamic economic system in which
technological innovation is the crucial input.

This study seeks to combine approaches from several different historical subdisciplines. It is a history of science and technology, but
not one that focuses on major figures like Thomas Edison, Ira Rem-
son, Henry Rowland, the Wright brothers, or others who lived during
the same period in which the Louisiana sugar industry was transformed in tems of scale and operations. Rather, it is a story of sec-
ond- and third-level scientists and engineers who are generally unknown to historians but nevertheless made valuable contributions
to the development of modern culture. If we are to truly understand
the impact of science upon society, and vice versa, we must study
this group of experts.

The story of the Louisiana sugar industry extends the traditional
boundaries of southern history. While the study characterizes manufacturing in the state, it places that activity in a broader, international context. Further, this book is a contribution to the literature related to the history of sugar, an area in which numerous
scholars have been working for generations.[10] Yet it is much more

10. The literature in this area is voluminous. In addition to Sitterson's study, important works include Sidney W. Mintz, *Sweetness and Power: The Place of Sugar
in Modern History* (New York, 1985); Noel Deerr, *The History of Sugar* (London,
1949–50); W. R. Aykroyd, *Sweet Malefactor* (London, 1967); Mark Schmitz, *Economic Analysis of Antebellum Sugar Plantations in Louisiana* (New York, 1977);
E. Ellis, *An Introduction to the History of Sugar as a Commodity* (Philadelphia,
1905); Edward O. von Lippmann, *Geschichte des Zuckers* (2 vols.; Berlin, 1929);
William Reed, *The History of Sugar and Sugar Yielding Plants* (London, 1866);
Peter L. Eisenberg, *The Sugar Industry in Pernambuco: Modernization Without
Change, 1840–1910* (Berkeley, 1974); Alfred S. Eichner, *The Emergence of Oligopoly:
Sugar Refining as a Case Study* (Baltimore, 1969); Francisco A. Scarano, *Sugar and*

than a work in this area because the history of the sugar industry in Louisiana illustrates themes in areas as diverse as the history of the professions and the role of the elite in promoting economic change. By integrating these various ways at looking at the past, I intend to raise questions that other scholars may find useful in their own areas of interest. Above all, I shall try to demonstrate the value and power of the history of science and technology in understanding one important episode in the development of American civilization.

Slavery in Puerto Rico: The Plantation Economy of Ponce, 1800–1850 (Madison, Wis., 1984).

1

The French Connection

Boom times are nothing new to the inhabitants of south Louisiana. New Orleans, Baton Rouge, Lafayette, New Iberia, Morgan City, and other nearby communities experienced spectacular growth during the late 1970s and early 1980s as the state's "oil patch" was systematically explored and exploited. With this surge of prosperity came the fast-food chains, housing developments, shopping malls, heliports, and metal prefabs—all testimony to economic expansion. Rusted out cars with Ohio, Michigan, and Pennsylvania tags reflect the migration of families from the depressed northern states to this promised land.

Yet amid the bustle of this plastic world stands a legacy from the past—plantation homes and sugar factories—that serves as a reminder of a similar rush of settlers and fortune hunters to the region some 160 years ago.[1] During the 1820s word spread that great profits could be made from planting sugarcane. The dream of riches lured both footloose fortune hunters and entire families from New England, New York, Virginia, North Carolina, Ireland, and elsewhere to this lush, semitropical, and in many ways mysterious, region.

The antebellum sugar bowl consisted of three chief areas within south Louisiana. The oldest and most established area extended

1. Clarence John Laughlin, *Ghosts Along the Mississippi* (New York, 1961).

nearly 130 miles north of New Orleans along the banks of the Mississippi River (including territory in Avoyelles, West Feliciana, Pointe Coupee, East and West Baton Rouge, Iberville, Assumption, Ascension, St. James, St. John the Baptist, St. Charles, and Jefferson parishes) and approximately 60 miles downriver into St. Bernard and Plaquemines parishes. A second major area of sugar cultivation and production developed along the banks of Bayou Lafourche southwest of New Orleans in Lafourche and Assumption parishes, including the towns of Donaldsonville and Thibodaux. Finally, a third region, the destination of many settlers during the rush of the 1820s, centered around the towns of New Iberia and Franklin along the banks of Bayou Teche and environs in Lafayette, St. Martin, Iberia, and St. Mary parishes. Other secondary areas, including Rapides and Terrebonne parishes, were also important sugar-producing areas, but the large plantations of the wealthy sugar barons that dominated the industry were almost invariably located in one of the three major geographical regions. It was there that nineteenth-century risk takers, in part motivated by favorable tariffs, the introduction of new cane varieties, and falling cotton prices, were willing to invest tens of thousands of dollars to purchase land, equipment, and slaves.[2] More often than not the harsh realities of freezes, floods, and labor problems shattered the overly optimistic dreams of these pioneer sugar planters, a situation not unlike that of many transplanted Yankees who came to the state during the 1980s.

The origins of the sugar bonanza can be traced to experiments in the cultivation and production of cane that initially took place during the 1790s. Until that time indigo was the chief cash crop of Louisiana, but the ravages of insects and the appearance of plant diseases caused local planters to look for alternatives. As early as the 1750s and 1760s some sugarcane indigenous to Santo Domingo was planted near New Orleans, but these tentative trials ended with Spanish cession and the subsequent revolt. In the early 1790s interest in sugar was revived.

As a result of the efforts of Joseph Solis, Antonio Mendez, Antoine Morin, and Etienne Boré, the profitability of sugar making in Louisi-

---

2. Sitterson, *Sugar Country,* 23–25; Timothy Flint, *Recollections of the Last Ten Years* (1826; rpr. New York, 1968), 325.

ana was demonstrated.[3] Solis had come to Louisiana from Cuba during the 1780s, or perhaps earlier, and along with his father established a plantation and erected a small grinding mill and rum distillery in Terre-aux-Boeufs in what is now St. Bernard Parish. In 1794 Solis sold his property and equipment for six thousand silver pesos to Antonio Mendez, a native of Havana. It was probably late in 1794 that Mendez hired Antoine Morin, a chemist and sugar maker, originally from Santo Domingo, to handle his crop. For reasons unknown, Morin was soon employed by planter Etienne Boré, who in 1795 made a crop of sugar that sold for twelve thousand dollars and who, if we are to accept the dramatic reconstruction of Boré's experiments by his grandson Charles Gayarré, convincingly demonstrated that syrup from Louisiana canes would granulate and produce raw sugar.

> Boré's attempt had excited the keenest interest; many had frequently visited him during the year 1795 to witness his preparations; gloomy predictions had been set afloat, and on the day when the grinding of the cane was to begin, a large number of the most respectable inhabitants had gathered in and about the sugar-house to be present at the failure or success of the experiment. . . . Suddenly the sugarmaker cried out with exultation, "It granulates!" . . . Each one of the by-standers pressed forward to ascertain the fact on the evidence of his own senses, and when it could no longer be doubted, there came a shout of joy, and all flocked around Etienne de Boré, overwhelming him with congratulations, and almost hugging the man they had called their savior—the savior of Louisiana.

Production of sugar steadily increased during the first quarter of the nineteenth century, and by 1830 over 33,000 tons of sugar were made in the state.[4]

The manufacture of sugar in Louisiana in the 1830s generally differed only slightly in method from the manufacture of sugar in the French, Spanish, and English Caribbean colonies of the eighteenth century. The process of converting sugarcane to raw sugar can be divided into five steps: grinding the cane; defecating and purifying the

3. Rene J. Le Gardeur, Jr., "The Origins of the Sugar Industry in Louisiana," in Center for Louisiana Studies, *Green Fields: Two Hundred Years of Louisiana Sugar* (Lafayette, La., 1980), 1–28.

4. Charles Gayarré, "A Louisiana Sugar Plantation of the Old Regime," *Harper's New Monthly Magazine*, LXXIV (1887), 607; A. Bouchereau, *Bouchereau's Revised Directory for 1917* (New Orleans, 1917), 38.

extracted juice; evaporating the juice to a viscous syrup; granulating the syrup, with the formation of sugar crystals; and potting, the separation of crystals and molasses by slow drainage.[5]

The sugarcane, normally harvested between late October and December, was taken in carts from the field to the sugarhouse. During the 1830s two methods of grinding the cane were practiced, one using animal power and the other employing the steam engine as the prime mover. In either case, the mill was elevated, not only to allow the juice to run into vats, but also to leave room to receive the exhausted cane, or bagasse. The animal-powered mill was usually of a vertical design, consisting of three fluted, interlocking cylinders, each 30 to 40 inches long and 20 to 25 inches in diameter. The oxen, horses, or mules applied power to the central cylinder of the mill by means of a lever or beam arrangement. Adjustments made to the distance between cylinders, in response to changes in the nature of the cane, were achieved by a system of cross keys and wedges.

By the 1830s steam engines began to supplant animal power. In 1828 only 120 of the 691 sugarhouses in Louisiana employed steam-powered mills; sixteen years later, 408 of the 762 mills employed the steam engine to crush the cane.[6] The steam-powered mill usually had its cylinders arranged in a horizontal configuration; these cylinders, 4 to 5½ feet long and 25 to 27 inches in diameter, were arranged triangularly, with one cylinder above and two beneath. The cane, in 3- to 4½-foot lengths, was brought to the mill by a cane carrier, a continuous chain that was powered by the steam engine. The carrier dropped the cane into a wooden hopper, and gravity carried it to the mill rollers. After passing through the mill, the cane was conveyed outside the sugarhouse by means of an inclined plane.

The mill was a constant source of anxiety for the planter. In his business, time was crucial; an early frost could lead to financial ruin. The mill's operation was uncertain; for example, if the cane piled up

5. Benjamin Silliman, *Manual on the Cultivation of the Sugar Cane and the Fabrication and Refinement of Sugar* (Washington, D.C., 1833), 30–41; H. O. Ames, *H. O. Ames' Improved Method of Evaporating Saccharine Juices by Steam* (New Orleans, 1859), 2–4; V. Alton Moody, *Slavery on Louisiana Sugar Plantations* (N.p. [1924?]), 48–52; Sitterson, *Sugar Country*, 137–44.

6. Silliman, *Manual on the Cultivation of the Sugar Cane*, 31; J. A. Leon, *On Sugar Cultivation* (London, 1848), 4.

at the entrance between the first two rollers, the roller shafts often broke suddenly. And the necessary parts and repair service were to be found only in New Orleans. Also, the ridges on the cane shafts—the nodes or joints—determined the distance set between the mill cylinders. Because of the varying distances, pressure was often inadequate to squeeze all the juice from the cane. Thus, a great deal of juice was often left in the cane after its passage through the mill.

The freshly extracted cane juice flowed through a spout into two large vats, or juice boxes, made of cypress planks. These vats, which had a capacity of several hundred gallons, were located in the mill room (sometimes called the laboratory), in an area separated from the boilers. Prior to boiling, a strainer was used to skim the juice contained in these boxes, thus removing scum, trash, and cane pieces. The juice was boiled in large open kettles that were set in brickwork and arranged over a furnace. This set of kettles was called a "battery," "equipage," or "kettle train." Each kettle was of a different capacity—the largest held from three hundred to five hundred gallons, and the smallest held seventy to one hundred gallons. The first and largest kettle was called the *grande*. The second was the *flambeau*, so called because the point of the flame touched the kettle. In the third kettle, or *sirop*, the juice was boiled down to the density of syrup. Finally, in the *batterie*, or last kettle, the concentrated syrup was "struck"; small, almost invisible crystals of sugar formed. At a number of plantations, a fifth kettle was placed between the sirop and flambeau. This fifth kettle was called the *proper clear* because it was in it that the juice began to become transparent.

At the beginning of the clarification and evaporation step, the contents of the nearly full grande were brought close to the boiling point, and the proper defecating agent, usually lime, was carefully added. Rising to the surface, the resultant scum was ladled off and discarded. The water in the juice evaporated, and the resulting sugar syrup was concentrated during this step. As the water evaporated, the juice was gradually ladled by hand into the flambeau. The juice, tempered again with lime in the flambeau, was eventually transferred to the sirop, and as it approached the proper density and quality, was placed in the batterie. The syrup, having reached its striking point in the batterie, was promptly turned out into the cooler. The last kettle was then recharged with a fresh supply from the sirop, and the sirop replenished by the contents of the flambeau, which in turn was recharged with the material from the grande. The transfer of juice from one kettle to another involved the danger of exposing a nearly empty kettle to the direct heat of the flame. The remaining liquor would then be carbonized, producing a lower quality sugar.

In the last kettle, or batterie, the sugar was granulated or "struck." The sugar maker used a number of criteria to determine the proper moment for crystallization. A common practice was to thrust a wooden-handled copper spoon into the batterie. If the syrup had a grained appearance and formed a slowly draining film over the entire spoon, it was ready for removal to the cooler. Another method of testing the syrup was to place a small amount of the boiled juices between the thumb and forefinger; when drawn out into a thread, properly boiled syrup would break dry and rise in a spiral.

The resulting concentrated syrup was placed in coolers, where it remained for about twenty-four hours. The crystallized mass was then dug out and placed in hogsheads (large casks) located in the draining room. The modern sugarhouses of the 1830s normally had two draining houses, located at right angles to the sugarhouse with connecting doors to the area occupied by the coolers. The floor consisted of small beams running crosswise and placed about eighteen inches apart. The hogsheads were placed on these beams, and underneath were placed the molasses cisterns, each covering an area of about twenty square feet and approximately sixteen to twenty inches deep. The molasses drained from three or four holes located in the bottom of the hogshead. To facilitate drainage, canes were often loosely stuck into these holes, and as the separation of crystallized sugar from molasses took place, the liquor drained down the sides of the canes into the cisterns. The by-product molasses was then set aside, later to be packed in barrels and sent to market.

The New Orleans levee was the hub of commercial activity, a place where hundreds of hogsheads containing raw sugar were loaded on steamboats and ocean vessels. Approximately one-half of the sugar and molasses made in Louisiana was sold to western customers.[7]

7. J. Carlyle Sitterson, "Financing and Marketing the Sugar Crop of the Old South," *Journal of Southern History*, X (1944), 188–90; Merl E. Reed, "Footnote to the Coast-

With the exception of a small quantity consumed in the local region, the remainder of the product was shipped to the east coast cities of Boston, New York, Baltimore, Savannah, or Charleston. Planters along Bayou Teche, having no convenient transport connection with New Orleans, frequently transported their sugar and molasses directly to eastern ports, even after the completion in 1855 of the New Orleans, Opelousas and Great Western Railroad between the Mississippi River and Berwick Bay.

During the 1830s an increasing number of Louisiana planters recognized the inefficiencies of the traditional open kettle method, both in terms of labor and fuel costs. It was necessary to station a worker at every kettle, not only for the "skipping out" of the juice by the attendants using long-handled scoops, but also for "brushing off" the impurities that rose to the surface. With the exception of the cost of the cane itself, cordwood to fuel the furnaces was the most expensive cost element of the sugar-making process. In addition, sugar making was an empirical art, and sometimes even the most experienced sugar boiler could not control the process enough to produce good quality, crystallized sugar. Planters gradually became aware of enormous losses of sugar during the process. Also, concern was growing among Louisiana planters over the quality of the product—particularly because Louisiana's raw sugar had a dark brown color and brought lower market prices than the lighter-colored raw sugar made in the Caribbean.

To improve their processes and thus their products, Louisiana planters looked to France for scientific and technical knowledge and expertise. Indeed, it seems that by the early 1830s the entire cane-sugar industry began to recognize the potentialities of scientific technology. The progressive spirit embraced by the typically tradition-bound planters resulted in part from the abolition of slavery in certain sugar-producing areas and in part from a phenomenal upsurge in sugar consumption by European working classes and the English in particular during the 1830s. As Sidney Mintz ably demonstrated in *Sweetness and Power*, increased demand during this period came

wise Trade—Some Teche Planters and Their Atlantic Factors," *Louisiana History*, VIII (1967), 191–97.

from the lower classes of society who were experiencing a transformation in diet as well as in economic environment.[8]

Beginning with the 1830s, new scientific techniques, manufacturing processes, and large-scale apparatuses transformed the manufacture of sugar on numerous plantations. These innovations, brought about in part through efforts of American inventors and chemists, were largely due to the diffusion of French chemical methods and beet-sugar technology into Louisiana. The reception and subsequent selective adoption of French scientific ideas by Louisiana planters was facilitated by the presence of a large population of French descendants and French-speaking American citizens residing on plantations located along the Mississippi River and Bayou Lafourche.

Frederick Law Olmsted, a frequent traveler to the southern states during the 1850s, noticed the influence of French scientific ideas on the progressive Louisiana planter community. He wrote in *A Journey in the Seaboard Slave States, With Remarks on their Economy* (1856):

> The whole process of sugar manufacturing, although chemical analysis proves that a large amount of saccharine is still wasted, has been within a few years greatly improved, principally by reason of the experiments and discoveries of the French chemists, whose labors have been directed by the purpose to lessen the cost of beet sugar. Apparatus for various processes in the manufacture, which they have invented or recommended, has been improved, and brought into a practical operation on a large scale on some of the Louisiana plantations, the owners of which are among the most intelligent, enterprising, and wealthy men of business in the United States.[9]

At least four pathways can be reconstructed for the diffusion of French science and technology into Louisiana. First, recent immigrants or Americans educated at French universities played a significant role in the injection of scientific knowledge into the Louisiana sugar industry. Second, the continental travels of planters and their employees often led to the acquisition of valuable practical knowl-

8. Mintz, *Sweetness and Power*, 160–202.
9. Frederick Law Olmsted, *A Journey in the Seaboard Slave States, With Remarks on Their Economy* (1856; rpr. New York, 1968), 670.

edge and theoretical principles. Third, in 1847 a study of sugar chemistry and manufacturing methods, sponsored by the United States Treasury, was made available to planters. This treatise described the numerous French contributions in both sugar chemistry and technology. Finally, the appearance of *DeBow's Review* in 1846 provided Louisiana planters with numerous essays on the cultivation and manufacture of sugar in Louisiana, as well as reprinted English and French articles and treatises on the scientific and engineering aspects of sugar manufacture.

French-educated scientists played a central role in improving the Louisiana sugar industry prior to the Civil War. J. B. Avequin, a New Orleans apothecary and pupil of Nicolas Louis Vauquelin, studied the chemistry of sugarcane and employed chemical analysis in determining the efficiency of milling during the 1830s and 1840s. Avequin apparently first became interested in sugar chemistry while in Santo Domingo during the early 1830s. He arrived in St. James Parish in 1832 where he analyzed the composition of sugarcane using extraction and distillation techniques. He remained active in this field until his death in 1861.[10] Norbert Rillieux, born a free quadroon (one-fourth Negro) in Louisiana in 1806, was a graduate of the Ecole Polytechnique.[11] As a student in Paris, Rillieux was interested in steam

10. J. B. Avequin, "Notice historique sur l'introduction des variétés de la canne à sucre à Louisiana, et analyse comparée," *Journal de Chimie Medicale,* II (1836), 26–34, 132–35; "Sur la matière cireuse de la canne à sucre," *Annales de Chimie et de Physique,* LXXV (1840), 218–22; "De la formation du cal dans les chaudières pendant la fabrication du sucre brut (sucre de canne)," *Journal de Pharmacie et de Chimie,* XXVII (1840), 15–20; "Du sucre d'erable aux Etats-Unis," *Journal de Pharmacie et de Chimie,* XXXII (1857), 280–85; "Des ennemies de la canne à sucre ou les insectes qui attaquent la canne à sucre dans les Antilles et en Louisiana," *Journal de Pharmacie et de Chimie,* XXXII (1857), 335–37; "La canne à sucre à la Louisiana," *Journal de Pharmacie et de Chimie,* XXXII (1857), 337–45. For Avequin's obituary, see New Orleans *Daily Picayune,* February 26, 1861, p. 6.

11. David Harbison, *Reaching for Freedom: Paul Cuffe, Norbert Rillieux, Ira Aldridge, James McCune Smith* (N.p., 1972); Harry A. Ploski and Warren Marr (eds.), *The Negro Almanac* (New York, 1976), 794–95; "Norbert Rillieux," *The Louisiana Planter and Sugar Manufacturer* (hereinafter cited as *LP*), XIII (1894), 285; P. Horsin-Deon, "Norbert Rillieux," *LP,* XIII (1894), 331; Marcus Christian, *Negro Ironworkers in Louisiana 1718–1900* (Gretna, La., 1972), 23–24; R. L. Desdunes, *Nos Hommes et Notre Histoire* (Montreal, 1911), 101–103; H. E. Sterkx, *The Free Negro in Ante-Bellum Louisiana* (Rutherford, N.J., 1972), 227–28.

engines, and he later published a series of papers on their use. During his residence in France, he became interested in the application of the steam engine to sugar manufacturing, but after having little success in selling his ideas to French refiners, Rillieux returned to Louisiana. There he began to experiment with vacuum evaporation and for over a decade conducted a series of trials that initially ended in failure. Finally succeeding, he patented his multiple-effect evaporation system in 1843 and further improved his invention with an 1845 patent. Rillieux's design was the premier engineering achievement in nineteenth-century sugar technology.

Antoine M. F. Chevet immigrated from France to Louisiana in 1839. Although Chevet was president of Jefferson College, located in St. James Parish, he was also an inventor. In 1847 he was awarded two patents for improvements in the manufacture of sugar.[12] Both innovations focused on improving the defecation step of the process. One method used high pressure to destroy impurities; the other employed a lime-sugar compound, rather than lime alone, as the defecating agent.

Pierre Adolph Rost was born in 1797 in France and was educated at the Ecole Polytechnique.[13] Having immigrated to New Orleans in 1816, he became a judge of the Supreme Court of Louisiana in 1838. A key figure in the Louisiana Agriculturists' and Mechanics' Association, he was described by one traveler in 1849 as "one of the few planters who study science to apply it to practical operations of planting sugar cane."[14] Thus, immigrants helped inject French science into Louisiana. In particular, the works of Avequin and Rillieux, the former in sugar chemistry, the latter in engineering, were central to the emergence of an improved sugar industry in the 1840s.

A number of planters learned of French innovations in their travels.

12. A. M. F. Chevet, "Improvement in the Manufacture of Sugar," U.S. Patent 4,985, February 27, 1847. See also Chevet, "Improvement in Sugar Making," U.S. Patent 5,276, September 4, 1847. For information on Chevet, See Chevet to Bannon G. Thibodeaux, July 12, 1846, U.S. Patent 4,985. These documents are all in Record Group 241, Records of the Patent Office, National Archives, Washington, D.C. (hereinafter cited as NA).

13. *National Cyclopedia of American Biography*, XI, 468.

14. Herbert Anthony Kellar (ed.), *Solon Robinson: Pioneer and Agriculturist, Selected Writings* (1936; rpr. New York, 1968), II, 168.

Valcour Aime, a St. James planter, and his neighbor, Pierre Michel La Pice, observed an improved method of manufacture (the Derosne and Cail apparatus) in their 1845 travels to Cuba and actively promoted it upon returning home.[15] La Pice was born on the island of Santo Domingo in 1797. After the massacre of whites on that island, the family settled in New Orleans in 1804. As a young man he entered into a mercantile partnership with a Mr. Millandon and represented the firm's interests in Natchez. In 1834 La Pice settled in St. James Parish and became a neighbor of Aime. There on his plantation hè made white sugar using charcoal filters and was the first to install a five-roller mill and bagasse burner in Louisiana.

Aime, born into an old Louisiana family in St. Charles Parish in 1798, acquired a sugar plantation as a young man in St. James Parish and developed the first sugar refinery in the state. As early as 1829 he employed steam machinery and adopted new boiling equipment, including Rillieux's multiple-effect evaporation system, as soon as it became available. Aime learned of developments in France by sending "a young man who had, under my tuition, become a pretty good sugar boiler, to take further lessons from a refiner living near St. Quentin. He has orders also to go to England, and to visit, before he returns, the refineries of our northern cities."[16]

Of all the information acquired by travelers to France during the 1840s, Judah P. Benjamin's observations and experiences were most widely disseminated to Louisiana planters. While his contemporaries recognized Benjamin as a knowledgeable and successful sugar planter in Louisiana, historians have focused on his career as secretary of state during the Civil War. Born in 1811 on the island of St. Thomas, Benjamin was educated at Yale and practiced law in New Orleans during the 1830s.[17] He subsequently purchased a sugar plan-

15. *Dictionary of American Biography*, I, 130; Valcour Aime, *Plantation Diary of the Late Mr. Valcour Aime* (New Orleans, 1878); *Biographical and Historical Memoirs of Louisiana* (Chicago, 1892), II, 533–34; Roulhac B. Toledano, "Louisiana's Golden Age: Valcour Aime in St. James Parish," *Louisiana History*, X (1969), 211–24.

16. "Sugar Culture and Manufacture of Louisiana and West Indies," *DeBow's Review*, IV (1847), 383.

17. Pierce Butler, *Judah P. Benjamin* (Philadelphia, 1906); Simon I. Neiman, *Judah Benjamin* (Indianapolis, 1963), 44–47; Robert Donthat Meade, *Judah P. Benjamin* (New York, 1943), 57–61; *National Cyclopedia of American Biography*, IV, 285.

tation. Before Benjamin lost the property through an unfavorable financial settlement during the late 1840s, he wrote the most perceptive essays of any Louisiana planter about sugar chemistry and new process machinery. His visits to Paris had resulted in "notes . . . of value, comprising references to what I have read, and statements of what I have seen and heard and subject so interesting to me as the sugar culture and manufacture, and have collected every thing of any value that has ever been published on the subject."[18] Benjamin contributed a number of articles to *DeBow's Review* about new apparatuses, the chemistry of sugar manufacturing, and advanced methods of chemical analysis.

A third way in which French ideas were introduced into Louisiana was in the widespread distribution of an 1847 treatise by R. S. McCulloch entitled *Report of Scientific Investigations Relative to the Chemical Nature of Saccharine Substances, and the Act of Manufacturing Sugar.* Richard Sears McCulloch was born in Baltimore in 1818.[19] While working as a laboratory assistant to Joseph Henry, McCulloch graduated with a bachelor of arts degree from the College of New Jersey in 1836. Acting on Henry's suggestion, he spent the winter of 1838–1839 in the study of practical chemistry at the laboratory of James Curtis Booth in Philadelphia. During the late 1830s, McCulloch was first employed by the Chesapeake and Ohio Canal Company, and later he was an observer in Girard College's Magnetic Observatory in Philadelphia. In 1840 he became a professor of natural philosophy, mechanics, and chemistry at Jefferson College in Canonsburg, Pennsylvania. McCulloch was appointed melter and refiner at the United States Mint in Philadelphia in 1846, and while he was in this position, he assisted the secretary of the treasury in studying chemical and manufacturing problems related to the sugar industry. In addition to his work on the chemistry and manufacture of sugar, McCulloch published a number of reports

18. "Sugar—Its Cultivation, Manufacture and Commerce, No. II," *DeBow's Review,* IV (1847), 297.

19. McCulloch's treatise is contained in A. D. Bache and R. S. McCulloch, *Reports for the Secretary of Treasury of Scientific Investigations in Relation to Sugar and Hydrometers* (Washington, D.C., 1848). Biographical information on McCulloch is from Milton Halsey Thomas, "Professor McCulloh of Princeton, Columbia and Points South," *Princeton University Library Chronicle,* IV (1947), 17–29.

on related subjects during the mid-1840s, including *A Report of Chemical Analyses of Sugars, Molasses, Etc.*, and *of Researches on Hydrometers* (February 21, 1845), and *A Report of Researches on Hydrometers and Spiritous Liquors* (June 2, 1848).

McCulloch's work had two purposes: first, to prevent revenue frauds by determining which analytical chemical procedures were the most specific and accurate for the analysis of sugar in molasses; second, to characterize the current status of scientific sugar manufacturing so that an assessment could be made of the relative merits of the methods practiced in Louisiana and the Caribbean. McCulloch's treatise on sugar manufacturing included a report of his own studies and presented a collection of chemical and technical treatises and articles translated from French. Indeed, this study was grounded on theoretical and applied studies conducted by a host of French or French-related investigators, including J. B. Avequin, Norbert Rillieux, Anselme Payen, Jacques François Dutrône la Couture, E. Degrand, Jean Baptiste Boussingault, Eugène Péclet, Jean Baptiste Dumas, Nicolas Vauquelin, T. Clerget, Henry Braconnot, Augustin Pierre Dubrunfaut, and Eugène Peligot.[20] For McCulloch, the modernization of the cane-sugar industry in Louisiana meant the injection of French ideas concerning the manufacture of beet sugar into an industry that was dominated by rule-of-thumb methods. McCulloch acknowledged that the European beet-sugar industry had benefited from the application of scientific innovations. In contrast, he felt that cane-sugar manufacturing had remained technically backward because the process was controlled by Negro slaves instead of scientifically trained experts. He wrote:

20. Among the treatises employed in McCulloch's report were: Jean Baptiste Dumas, *Traité de Chimie Appliquée Aux Arts* (Paris, 1828–46); Eugène Péclet, *Traité de la chaleur consideree dans ses applications* (Paris, 1843); Jean Baptiste Boussingault, *Rural economy, in its relations with chemistry, physics and meteorology*, trans. George Law (New York, 1845); E. Degrand, *Fabrication et Raffinage du sucre: Notice sur la concentration des jus sucres et la cuisson des sirops* (Paris, 1835); Jacques François Dutrône la Couture, *Precis sur la canne et sur les moyens d'en extraire le sel essentiel* (Paris, 1790); Anselme Payen, *Cours de chimie elementaire et industrielle* (2 vols.; Paris, 1832–33); Eugène Peligot, *Recherches sur la nature et les proprietes chimiques des sucres* (Paris, 1838).

While the beet sugar industry has had great difficulties to contend with in the very impure nature of the juice employed, it yet has received in France every facility and advantage which the science, intelligence, and fostering care of a powerful and enterprising nation could afford. On the other hand, the cultivation of the cane, and the art of extracting sugar from it, seem almost to have been left for centuries to the management of the ignorant and stupid negro. . . . Thus, in these two similar and rival industries, we behold science overcoming obstacles imposed by nature, and contending with indolence favored in every respect.[21]

Although the role of immigrants, the observations from travelers, and government reports stimulated the diffusion of French science into Louisiana, the numerous articles, reprints, and opinions published in *DeBow's Review* provided the Louisiana planter with a steady supply of valuable scientific and technical information.[22] This magazine, founded by J. D. B. DeBow in 1846, contained numerous articles describing scientific sugar culture and manufacture; reprints of valuable treatises, including Evans' *A Handbook for Sugar Planters*; translations of French articles, including those by E. Degrand and Anselme Payen; reports and financial statements from Valcour Aime, Norbert Rillieux, and others; and discussions of new processes patented in the United States and Europe. DeBow himself commented from time to time on the development of the sugar in-

21. Bache and McCulloch, *Reports for the Secretary of Treasury*, 221.

22. Ottis Clark Skipper, *J. D. B. DeBow, Magazinist of the Old South* (Athens, Ga., 1958). There are numerous articles in *DeBow's Review* that attempted to join together theory and practice in the manufacture of sugar. They included: J. D. B. DeBow, "Agricultural Associations," I (1846), 161–68; J. P. Benjamin, "Louisiana Sugar," II (1846), 322–45; E. J. Forstall, "Louisiana Sugar," I (1846), 53–55; L. B. Stone, "On the Crystallization of Sugar," III (1847), 230–34; Maunsel White, "The University of Louisiana," III (1847), 260–65; L. B. Stone, "A Few Notes for Sugar Planters," III (1847), 297–305, 376–96; "Southern and Western Agricultural and Mechanics' Associations," IV (1847), 415–50; "McCulloh's Report on Sugar—Reviewed," V (1848), 244–57; "Avequin's Review of McCulloh, Reviewed," V (1848), 334–63; "Sugar Making in Louisiana," V (1848), 285–93; J. P. Benjamin, "Soleil's Saccharometer," V (1848), 357–64. DeBow also condensed important treatises including William J. Evans, *A Handbook for Planters* (London, 1847) in "Sugar—Its Cultivation, Manufacture and Commerce," IV (1847), 152–59, 296–310. See also J. D. B. DeBow, *The Industrial Resources of the Southern and Western States* (New Orleans, 1853).

dustry in Louisiana, usually within the context of related institutions—the University of Louisiana and the Louisiana Agriculturists' and Mechanics' Association. Judah P. Benjamin wrote a number of the longer articles on sugar. DeBow also acquired information from J. B. Avequin, Valcour Aime, Maunsel White, P. M. La Pice, Thomas Packwood, and other Louisiana planters. He obtained foreign reports from a number of sources, including R. S. McCulloch in Philadelphia, and Thomas G. Clemson, located at the United States consulate in Brussels.[23]

DeBow's Review played an important role in the planters' early attempts to modernize the Louisiana sugar industry. It not only contained articles describing French processes for manufacturing sugar, but also propounded a progressive view concerning the role of science and its relationship to the manufacturing arts, usually with reference to French science. It appears that several planters viewed science as both a guide and an active force for industrial improvement. One *DeBow's* article stated, "It is plainly important then, to gather from science the light necessary to protect us from the delusions of an excited imagination, and to guide us to the way of safe and profitable enterprise." In an article entitled "The Mission of Science," DeBow maintained that science had a divine purpose, and that scientific progress was the will of God. He felt that God fostered and perfected science, and that God "will sweep away the ignorance and prejudice that oppose barriers to its advance—that he will permit no more the great book of nature to be closed, . . . that a lustre, brighter each day, may beam from its pages, diffusing and reflecting . . . chasing away the clouds of error and idolatry, . . . and ushering in the great and desired year—the Millenium of Science." Others, like L. B. Stone of New Orleans, felt that the advancement of science would be due more to the efforts of man than those of God. He asserted that "we should despise the feeling that there is anything either in the simple or refined operations of nature too large for our comprehension, or too minute to be worthy [of] our devoted attention."[24]

This new ideology of science proclaimed that man had control

23. "Melsen's Sugar Manufacture," *DeBow's Review,* VIII (1850), 111–22.

24. L. B. Stone, "A Few Notes for Sugar Planters," *ibid.,* III (1847), 298; J. D. B. DeBow, "The Mission of Science," *ibid.,* XXIII (1857), 662–63; L. B. Stone, "On Crystallization of Sugar," *ibid.,* III (1847), 396.

over his circumstances and that, as a consequence of intelligent activity, improvements would eventually follow. Pierre Rost, in his 1845 address to the Agriculturists' and Mechanics' Association of Louisiana, stated: "The innate faculty of our people is to subdue the physical world. . . . With heroic determination, then, speed the plow; bear in mind that to go ahead without ever taking difficulties into account, and by that means to succeed when others dare not undertake, is emphatically the AMERICAN SYSTEM."[25]

Judah P. Benjamin, the New Orleans lawyer and Plaquemines Parish planter, wrote the most detailed essays in *DeBow's Review* concerning the relationship between science and sugar manufacture. It appears that his ideas about science were significantly influenced by his observations of French science and French beet-sugar manufacture. Benjamin felt that only in France "have the researches of men of science been so ardently and extensively directed to the practical application of the discoveries of the laboratory to the improvement of the manufacturing industry." He recalled that if an industry were to be transformed by the injection of science, the presence of experts familiar with fundamental physical laws would be necessary. He noted that in France

> the men who conduct their manufactories and refineries are in very many instances carefully educated with a view to this pursuit, and only enter into the practice of their art after being intimately acquainted with its theory in all its branches. With those principles of physics and mechanics which will enable them thoroughly to understand the working of the machinery employed, and with those discoveries of modern chemistry which can best enlighten them as to the real nature of the delicate and beautiful process. . . . The advantages possessed by such men, surrounded by all the means and appliances of advanced civilization, with ready reference on all subjects of doubt or difficulty to men of eminent scientific attainments, and with every facility for obtaining at the cheapest rate, the supply or repair of machinery and material of every kind, over the indolent or ignorant colonial Planter, or even over our own more intelligent agriculturists are inappreciable.[26]

25. "Southern and Western Agricultural and Mechanics' Associations," *ibid.*, IV (1847), 435–36.

26. "J. P. Benjamin's Address on Agriculture," *ibid.*, V (1848), 47.

Benjamin felt that the Louisiana industry was backward and would remain that way until planters became acquainted with the basic principles of chemistry. He considered Louisiana planters to be quite familiar with the mechanical principles associated with the steam engine, but felt that their knowledge of acids and bases was lacking. Benjamin also saw the value of cooperative investigation and problem solving. The Agriculturists' and Mechanics' Association of Louisiana was one institution that could bring planters, their problems, and their ideas together. The activities on Benjamin's Belle Chasse plantation, as recounted by one biographer, seemed to reflect his ideas concerning mutual cooperation. "And frequently, for quite long visits, came the dried up little chemist, Rillieux, always the centre of an admiring and interested group of planters as he explained this or that point in the chemistry of sugar or the working of his apparatus. . . . 'Bellechasse' became not only a sort of social focus for the planters of the neighborhood, but the scene of a symposium, as it were, on sugar."[27]

Planters like Benjamin, Aime, La Pice, and others recognized that science was the key to the improvement of traditional processes and to increased profits. It was natural that south Louisiana planters would be receptive to European ideas, since the region itself was a melting pot of capitalists having diverse origins and a land in which French and English were often interchangeably spoken. Yet this flow of ideas and methods had to be facilitated by institutions as well as individuals, and it was the weakness of local organizations that proved to be the greatest stumbling block to modernization in the years before the Civil War.

27. Butler, *Judah P. Benjamin*, 59.

2

Local Institutions and the Adoption of Scientific and Technical Innovations 1830–1861

One of the most fascinating aspects of the antebellum Louisiana sugar industry, which profited greatly from slave labor, was that several of its leaders, including Judah P. Benjamin, Valcour Aime, P. A. Rost, and P. M. La Pice were receptive to science-based technology. Perhaps this was so because the large sugar planter and manufacturer was not tied to the soil in the same way as the small grower; rather he was a protocapitalist with an international vision, only one generation away from the likes of an Andrew Carnegie. Recognizing that innovative ideas led to higher earnings and long-term stability, the progressive elite in Louisiana welcomed scientific and technological change. Aime's and Benjamin's fascination with French science cannot be seen simply as dilettantism; rather, it was a reflection of an attitude on the part of the planter elite that science was power. How to harness that power and use it to sustain the growth of the sugar industry in Louisiana was a crucial problem during the antebellum period and one that ultimately was hindered by several institutional obstacles.

Prior to the 1830s, chemical knowledge was employed rarely, if at all, on the Louisiana sugar plantation. The planter, or overseer, in charge of the sugarhouse used rule-of-thumb tests, experiential knowledge, and visual inspection to determine whether or not the sugar-making process was being properly conducted. One observer remarked in 1831 that "it was the eye alone that was to determine the ripeness of cane, and it was the eye and touch alone that were to determine the point at which the syrup was sufficiently boiled to granulate, experience and individual judgement were alone relied on, and we know how variable, and how little faith can generally be placed in either."[1]

Yet, between 1830 and 1850 progressive French ideas did find their way into the sugar industry of Louisiana. French chemical and analytical techniques proved to be useful not only in evaluating agricultural practices but also in controlling the manufacturing processes of the sugarhouse. In a few isolated instances, chemists determined the composition of the sugarcane, the extracted juices, the ashes of bagasse, and the constituents of evaporation scale. Also, systematic experimentation was employed to trace the losses of sugar during processing, as well as in monitoring and controlling the defecation step.

J. B. Avequin was perhaps the first scientist to apply French chemical knowledge to the Louisiana sugar industry. Avequin employed various separation techniques and methods of chemical analysis to numerous process problems. For example, Avequin studied the ash of bagasse, or cane that had passed through the mill, and discovered that it contained large percentages of silica, carbonate of lime, carbonate of potash, and oxide of iron.[2] In 1832 Avequin analyzed one hundred pounds of Otaheite and ribbon variety canes for water, saccharine matter, fiber, and salts and found that while the Otaheite canes contained about 1 percent more sugar, the ribbon variety had more fiber and salt.[3] The apothecary also isolated and identified numerous natural products contained in trace quantities in the cane, and he found seventeen different components in freshly extracted juices, one of which was uncrystallizable sugar and glucose.[4] One of Avequin's last published papers was a discussion of his isolation of a waxlike natural product coating the cane, which he called *cerosie*.

1. *House Documents*, 21st Cong., 2nd Sess., No. 62, p. 47.
2. Silliman, *Manual on the Cultivation of the Sugar Cane*, 29.
3. Leon, *On Sugar Cultivation*, 7.
4. M. Avequin, "Sur la Matière cireuse de la canne à sucre," *Annales de Chimie et de Physique*, LXXX (1840), 218–22; Eugène Peligot, *Recherches sur la Composition Chimique de la Canne a Sucre de la Martinique* (Paris, 1839), 30.

His work attracted the attention of the most notable French chemist of the day, Jean Baptiste Dumas, who not only conducted an elemental analysis of this new material, but also included Avequin's chemical contributions in his *Traite de Chimie Appliquee aux Arts.*[5]

Avequin's analyses were often conducted with the intention of solving manufacturing problems. For example, he analyzed the mineral scale forming on the steam tubes of vacuum pans.[6] Most significantly, he conducted numerous studies of mill extraction efficiencies. By running a series of experiments with both Otaheite and ribbon cane varieties, Avequin determined that passing 1,000 parts of Otaheite cane through a West Point foundry mill yielded 560 parts of juice and 440 of bagasse. In contrast, similar treatment of 1,000 parts of ribbon cane produced only 472 parts of juice and 528 of bagasse.[7] During the 1840s, Avequin continued his research on the relationship between varieties of cane and percentage of juice extraction.[8]

Since the methods employed by Avequin in his studies of manufacturing efficiency required university training in chemistry, only a few Louisiana planters could even hope to understand the scientific principles that formed the basis of his work. However, the application of simple methods of chemical control, particularly the use of the hydrometer, was well within the capability of the intelligent planter, and this instrument was generally adopted before the Civil War. Baumé's hydrometer was used to determine the density of expressed cane juice. The observed density was then correlated to the percentage of sugar contained in solution, usually by means of a table. A number of planters employed this device in determining both the proper state of clarification and the correct density for the strike. One planter wrote optimistically in 1831 that "there is a small instrument, of little cost, and of great simplicity in its use, which measures with as much distinctiveness the quality of sweets contained in the juice of the cane."[9] Yet the hydrometer was only a

5. Dumas, *Traite de Chimie Appliquee Aux Arts,* VI, 218–28.

6. Bache and McCulloch, *Reports for the Secretary of Treasury,* 578.

7. Peligot, *Recherches sur la Composition Chimique de la Canne a Sucre de la Martinique,* 31.

8. *DeBow's Review,* VI (1848), 31.

9. *House Documents,* 21st Cong., 2nd Sess., No. 62, p. 43.

partial solution to the numerous process problems confronting the planters. It was an unspecific instrument, responding to total dissolved solids, not to sugar alone.

The planters' need to control the manufacturing process focused upon methods to determine the proper dosage of lime to be added in the defecation step. Although the hydrometer proved to be ineffective in providing the planter with this information, some sugar makers used empirical sampling tests to obtain the proper levels of the defecating agent. Spot tests and systematic trials in which small samples of cane juice were treated with known dosages of lime were practiced in some Louisiana sugarhouses by the mid-1830s. The appearance of flocculi in a clear glass would indicate the proper proportion of lime to be added to the kettles.[10] In the late 1840s, a similar method was outlined in *DeBow's Review*. The procedure, based on the work of the French chemist Anselme Payen, employed one gallon of juice and required the successive addition of known quantities of lime to the hot liquid. Filtration of a drawn sample and subsequent observation determined the proper tempering of the juice with lime. With assurance Payen asserted that "instead of spoiling entire strikes or batteries by different or excessive doses of lime, the manufacturer would proceed in perfect confidence."[11] Payen's method, however, assumed that the chemical composition of the extracted juice remained constant, and this condition could not be taken for granted in milling the immature cane grown in Louisiana. In order that the proper amount of lime and other materials could be added during the defecation step, an analytical procedure was needed that could quickly determine the subtle chemical changes that were taking place in freshly expressed saccharine material. The answer to this problem lay in the application of polariscopic techniques to the routine of the sugarhouse.

In contrast to the simplicity of the hydrometer, the polariscope was the most sophisticated analytical instrument introduced into Louisiana prior to the Civil War. Polarized light is characterized by vibrations confined to a single plane, and thus these rays are in-

10. Silliman, *Manual on the Cultivation of the Sugar Cane*, 35.
11. Judah P. Benjamin, "Louisiana Sugar," *DeBow's Review*, II (1846), 336.

capable of being reflected or refracted at certain angles.[12] During the first half of the nineteenth century, this type of radiation was most commonly obtained in the laboratory by passing light through a Nicol prism, which consisted of two transparent calcite crystals cemented together by Canada balsam. Since the calcite crystal had an optical axis for which the velocity of light is independent of its state of polarization, the crystal is transparent for only one polarization direction. Hence, light passing through the Nicol prism becomes linearly polarized. Certain organic molecules, including sucrose and glucose, interact with polarized light and rotate the plane of polarization. This deviation can be employed to determine the concentration of optically active substances in solution.

In 1840 the polariscope had been used exclusively for research purposes, but by the end of the decade its design was modified to meet the needs of the sugarhouse laboratory. Jean Baptiste Biot (1774–1862) employed a theoretically simple, but hard to use, polariscope in 1842 to determine the sugar content of cornstalks. His polariscopic apparatus included a light source, a polarizer (Nicol prism), a sample tube, an analyzer (another Nicol prism), and a telescope. Light from the source was passed through the polarizer, and these polarized rays were then passed through the test solution. Since both the polarizer and analyzer could be moved with respect to one another, Biot could make adjustments in order to create either maximum darkness or brightness. For example, with the sample tube empty and the analyzer set at ninety degrees with respect to the polarizer, no light was transmitted and the viewer saw a dark field through the telescope. When an optically active substance was placed in the sample tube, the light emanating from the polarizer was subsequently rotated, and light emerged from the analyzer. By rotating the analyzer to again establish the dark field, Biot could determine the rotation of the solution. The resultant angle of deviation was directly proportional to the length of the sample tube, the density of the sample, and the amount of optically active material in solution.

Biot's instrument and methods had several shortcomings, however. To begin with, the determination was based upon rotating the rela-

12. *McGraw-Hill Encyclopedia of Science and Technology* (1982 ed.), X, 592–97.

tive position of the Nicol prisms, and it was difficult to visually establish the point of total extinction. Also, Biot used sample tubes of various lengths, and in order to insure correct calculations, the measured density of each sample was required. In addition, polarimetric readings had to be conducted in a special room, since stray light interfered with the assay.

During the 1840s polariscopes were radically redesigned.[13] In 1846 Henri Soleil, a French optician and instrument maker, incorporated the basic features of Biot's polariscope in the design of a saccharimeter that enabled sugar refiners to conduct routine and rapid assays.[14] Soleil's apparatus was illuminated by daylight or other common light sources, and its features allowed the sugar chemist to read the percentage of sugar directly from a calibrated vernier scale. By placing a pair of carefully constructed quartz plates (the transition-tint plate) between the polarizer and the sample, Soleil constructed an instrument by which small changes in optical rotation could be easily measured.

Soleil's most important innovation was his addition of a quartz-wedge compensator between the sample tube and analyzer prism of the saccharimeter. The compensator consisted of a collection of positive and negative quartz wedges.[15] These elements could be moved relative to one another and thus provided an optical arrangement in which a dial scale could be utilized. After standardizing the compensator with one normal weight sucrose solution, the purity of any sample could be found by merely resetting the compensator and reading the percentage of sugar directly from the scale.

Apparently a few of Soleil's instruments were in use on Louisiana plantations before 1850. Judah P. Benjamin, in an 1848 article in

13. C. A. Browne, "Origin of the Clerget Method," *Chemical and Engineering News,* XX (1942), 322–24.

14. Henri Soleil, "Nouvel appareil propre à la mesure des deviations dans les expériences de polarisation rotatorie," *Comptes Rendus de l'Academie des Science,* XXI (1846), 426–30; M. T. Clerget, "Analyse Des Substances Sacchariferes," *Annales de Chimie et de Physique,* XXVI (1849), 175–207; Henri Soleil and M. Jules Duboseq, "Note sur un nouveau compensateur pour le saccharimetre," *Comptes Rendus de l'Academie des Science,* XXXI (1850), 248–50.

15. George William Rolfe, *The Polariscope in the Chemical Laboratory* (New York, 1905), 31–34.

DeBow's Review, described Soleil's design. Benjamin first characterized the nature of polarized light and the proper manipulations required for one to obtain a correct measurement. Further, he claimed that such an instrument was not beyond the abilities of an intelligent planter. "M. Soleil has succeeded in inventing a different instrument, which although depending on the same principles, may be used at all times and in all situations, without any necessity of calculation, and which can be practically applied by persons ignorant of the theory by which he was guided, in the same manner as the thermometer and barometer are daily used by persons unacquainted with the physical laws which influence the rise or fall of the mercury in their columns."[16] Benjamin went on to describe three different polariscopic assay procedures. One was a simple method to determine the percentage of crystallizable sugar in a solution that contained no other types of sugar. He also provided a detailed procedure for analysis of molasses. Benjamin concluded his essay with a step-by-step description of T. Clerget's method for analyzing the quantity of crystallizable sugar in the presence of noncrystallizable or inverted sugars.

Thus, new methods of analysis and new instruments were introduced into the Louisiana sugar industry during the 1830s and 1840s, but they were employed by a relatively small group of planters. In particular, the chemical manipulations and procedures that required extensive chemical knowledge were limited to the small number of individuals normally associated with the medical community. Even this group, with the exception of Avequin, was reluctant to extend its expertise beyond its traditional professional arena. One New Orleans physician, Samuel A. Cartwright, wrote: "Unfortunately, . . . for the South, if members of the medical profession interest themselves in matters of public utility, whether it be political economy, agriculture, manufactures or internal improvements of any description, the ignorant, indolent, envious and jealous, are always ready to injure and curtail their usefulness by sneering at them as dangerous experimenters, crack-brained theorizers, too learned for the practical duties of their profession, as if spending their leisure moments

16. Judah P. Benjamin, "Soleil's Saccharometer," *DeBow's Review*, V (1848), 357–64.

in the chemical laboratory, or at books or the writing-desk, would disqualify them for practice."[17] Physicians and pharmacists served the manufacturing community only in special isolated cases. Since there was an insufficient supply of local scientists necessary to sustain scientific innovation, chemical knowledge was never efficiently harnessed to address the cultivation and manufacturing problems associated with the antebellum Louisiana sugar industry.

While French chemical knowledge had a limited reception in Louisiana because of the lack of trained experts, French engineering innovations were more easily assimilated into this region already quite familiar with steam technology. In 1838 Louisiana was second only to Pennsylvania in the number of operating steam engines in the United States.[18] Planters were familiar with the steam engine and had attempted on numerous occasions during the 1830s to modify this technology by applying high and low pressure steam for improving the sugar manufacturing process. Even before the introduction of the French apparatus, Louisiana planters had experimented with steam technology. Benjamin Silliman wrote in 1833:

> A method of clarification by steam, is coming into use upon the larger and best conducted plantations. The sap is run off into wooden vats, lined with copper or lead, of the capacity of juice cisterns, . . . A copper steam pipe, between two and three inches in diameter, traverses, five or six times, the bottom of the cistern, to which it is firmly attached. The cane liquor is introduced, and when the pipes are covered with it, steam is let in, either from a boiler provided expressly for it, or from the boiler of the engine, or it is the escape steam which is employed.[19]

High and low pressure steam, conducted by tubes, was one method of attempting to control the heating of open kettles, thus making the clarification and evaporation processes more efficient. Overheating, a common problem in using kettles, did not take place with steam. The evaporation process could be easily and instantaneously con-

17. Dr. Cartwright, "Extension of the Sugar Region of the United States," *DeBow's Review,* XXIV (1853), 203.

18. Reynold M. Wik, *Steam Power on the American Farm* (Philadelphia, 1953), 6. An analysis of changes in the number of steam engines employed in the Louisiana sugar industry can be made for the period 1844 to 1860 by using the annual editions of P. A. Champomier, *Statement of the Sugar Crop Made in Louisiana.*

19. Silliman, *Manual on the Cultivation of the Sugar Cane,* 36.

trolled by the steam valve located next to the apparatus. By merely replacing the variability of the furnace fires with a boiler and a series of steam coils, the sugar boiler was able to have more control over the process.

Several designs of steam coils were employed in Louisiana, including one that consisted of a series of double concentric tubes. The steam was conveyed through the inner tubes and the condensate returned to the boilers by the outer tubes. Another system used one simple, continuous tube for both heating and condensation. In 1830 a number of Louisiana planters attempted to clarify and evaporate solely by steam. Unfortunately, this trial ended in failure and "shook for a long time the sugar planters of that most enterprising country."[20] Defects in construction, apparently stemming from the difficulty of joining narrow copper tubes together without leaks, limited the utility of these early designs.

In 1832 another device, the steam vacuum pan, was successfully used by Thomas Morgan of Plaquemines Parish. Other planters, including Valcour Aime, soon incorporated this design into their sugarhouses. Patented in 1813 by the Englishman Edward Charles Howard, the vacuum pan consisted of a closed, nearly hemispherical vessel. The pan had a double bottom and was heated by a serpentine steam coil. From the bottom of the apparatus, one pipe drew off the condensate while a larger one, passing through the double bottom, drew off the syrup from the chamber after it reached its proper concentration. Two pipes entered the top of the hemispherical pan—one supplied the dilute syrup to be concentrated; the other led to an air pump. Normally, a trap to collect entrained syrup was placed between the air pump and the pan, and a water condenser was often employed to liquefy the steam vapors originating in the syrup chamber. Both a thermometer and a barometer were attached to the apparatus. In addition, a "proofstick," or indented brass rod, passed into the interior of the pan and allowed the sugar maker to withdraw a sample without breaking the vacuum.

With the exception of the vacuum pan, attempted improvements to the sugar-making process initiated during the 1830s usually ended in frustration and failure. For example, between 1831 and 1835

20. Leon, *On Sugar Cultivation*, 4.

Norbert Rillieux found that Louisiana planters were not interested in employing his evaporation apparatus of advanced design. Rillieux was forced to erect the equipment at his own expense, and on one occasion he "was compelled to take it down before the trial, because it happened to be in the way and it was deemed absurd to believe that the cane juice of Louisiana could be boiled by steam."[21]

The frustration of the 1830s gave way to numerous successes during the 1840s. By the early 1840s planters were anxious to employ processes that produced a high-grade sugar, and workable French or French-inspired designs were now welcomed in Louisiana. By 1846 twenty-five to thirty Louisiana plantations had adopted one or more sophisticated methods of manufacture. Almost all of these sugarhouses used animal charcoal, or bone black, to decolorize the syrup. With the exception of the Morgan and Aime plantations, which were technologically several years ahead of the others, most of these changes took place between 1843 and 1846. The French Derosne and Cail evaporation apparatus was employed by Aime and by La Pice; eight plantations used the Rillieux multiple-effect design. The other planters used the Howard vacuum pan, except Edward Forstall, who installed steam coils in open kettles.

Why was this new technology so readily adopted in such a short span of time? One reason was that the price levels of both raw and white sugars were favorable for the expansion and modernization of the industry. With the exception of the years 1842 and 1843, raw sugar prices averaged over five cents per pound throughout the 1840s, and an even higher price level existed during the 1850s.[22] However, contemporary sugar planters felt that it was not price levels that transformed their industry; rather, it was the implementation of the tariff bill of 1842. James DeBow stated that the tariff of 1842 caused "to start up, as if by magic, . . . the costly mansion and the magnificent sugar mill."[23] Edward J. Forstall, in an 1845 pamphlet entitled *Agricultural Productions of Louisiana Embracing Valuable Information Relative to the Cotton, Sugar and Molasses Interests, and*

21. "Sugar," *DeBow's Review*, VII (1849), 58.

22. Lewis Cecil Gray, *History of Agriculture in the Southern States* (1933; rpr. Clifton, N.J., 1973), II, 744.

23. DeBow, *Industrial Resources*, III, 445.

the Effects Upon the Same of the Tariff of 1842, was more specific in making a causal relationship between the tariff and new technology.

> The Tariff of 1842 was their [Louisiana planters'] salvation; it at once restored confidence . . . , enabled the planter to improve his sugar works, to clear and drain his lands. . . . Thousands of Irishmen were soon seen digging canals in all directions; engineers putting up new engines, or repairing old ones—masons, setting sugar kettles on improved plans. . . . A steam apparatus, for the purpose of boiling in vacuo and producing white sugar direct from the cane, was put up last year . . . ; its success was such to induce another planter to order one. . . . Not less than five large estates will be working, this year, on the white sugar system.[24]

By restructuring the duty rates on raw and refined sugars, the tariff of 1842 conferred a significant financial advantage upon those planters employing either the vacuum pan or a multiple-effect evaporation system. While the product of open kettles was a brown, raw sugar, these new evaporators made a white, nearly pure sugar. In the previous tariff law, passed in 1832, only a small advantage was given to the manufacturer of a better grade product: a duty of $2\frac{1}{2}$ cents on raw sugar and $3\frac{1}{8}$ cents on clayed sugar (decolorized with mineral clay). The tariff of 1842 retained the $2\frac{1}{2}$-cent rate on raw sugar but increased the duty from $3\frac{1}{8}$ to 6 cents on clayed, white, and powdered sugars.[25] This law profoundly influenced the development of the Louisiana sugar industry. Concurrent with the maturation of French chemical and engineering knowledge in sugar manufacturing, the tariff of 1842 created an economic climate favorable for capital investment. Thus, the injection of French knowledge was catalyzed by the implementation of a favorable federal economic policy. However, the rates associated with the tariff of 1842 remained in effect only until 1846, when the duty was lowered to an ad valorem rate of 30 percent on both raw and refined sugars. Nevertheless, the adoption of new technology by Louisiana planters continued until the Civil War. By 1859 at least sixty-three plantations used either open steam pans, vacuum evaporators, or multiple-effect apparatuses. In particular, those planters who invested in advanced-design equip-

---

24. Edward J. Forstall, *Agricultural Productions of Louisiana* (N.p., 1845), 7.
25. Gray, *History of Agriculture in the Southern States to 1860*, II, 745.

ment during the 1840s were in a position to acquire more land and cultivate more cane during the boom times experienced by the Louisiana sugar industry during the 1850s.[26]

The emergence of new apparatuses in the 1840s was closely linked to innovations in railroad technology. It was no accident that the system of heating pipes employed by Rillieux was similar to the arrangement of pipes in the latest designs of railroad boilers. The two most innovative designs introduced in Louisiana during the 1840s were the Degrand apparatus, modified by Derosne and Cail, and the Rillieux multiple-effect evaporator. Both of these designs were based on the recovery of the latent heat of steam.

The Degrand apparatus was first described in an 1836 pamphlet issued by the *Journal des Connaissances Usuelles.*[27] It was subsequently modified by Louis-Charles Derosne and Jean-Francois Cail. Derosne was born in 1780 and studied pharmacy. After managing an apothecary shop in Paris, he began studying the refinement of raw sugar in 1808. He succeeded in preparing loaf sugar from the sugar beet the following year. He also studied the use of animal charcoal in 1813, and he invented a method of continuous distillation in 1817. By 1824 Derosne had established his association with Cail. As partners, they established a foundry for the manufacture of locomotives in 1839, and produced them in large quantities after 1844. Born the son of a cart maker in 1804, Cail was trained as a boilermaker.[28] After working on the gas-lighting system for the city of Paris, he joined Derosne. Like Derosne, Cail had constructed a continuous still, and in 1834 he designed a double-effect evaporator for the beet-sugar industry. After 1840 his work was totally occupied in the construction of locomotives.

The Derosne and Cail apparatus employed an oscillating steam engine as an air pump, which was then connected through a bent tube to a hemispherical pan heated by a steam coil. The water vapor from the syrup (heated in the closed pan) passed through the bent tube,

26. P. A. Champomier, *Statement of the Sugar Crop Made in Louisiana in 1859–1860* (New Orleans, 1860); Schmitz, *Economic Analysis of Antebellum Sugar Plantations*, 12–41.

27. *Fabrication et Raffinage du Sucre* (Paris, 1836).

28. *Dictionnaire de Biographie Francaise* (1956), VII, 840–41, *Dictionnaire de Biographie Francais* (1965), X, 1143.

where it condensed, transferring its heat to the dilute syrup pouring over the tube in fine streams. In this way, the dilute syrup was preheated and somewhat concentrated before entering the closed, hemispherical evaporation chamber. Moreover, the condensation of the vapor in the tube created additional vacuum.

Like the apparatus of Derosne and Cail, Rillieux's chief innovation also utilized the principle of latent heat. This equipment employed a series of connected chambers for the evaporation of syrups and it enabled the sugar maker to conserve even more fuel. The apparatus consisted of three or four closed evaporating pans interconnected by a succession of steam chambers. Steam resulting from the evaporation of the syrup was conducted from one vessel to the next. The second pan, operating at a lower pressure than the preceding one, was heated by the vapor of the first and in turn furnished the heat for the third pan, which was operated at still lower pressure. The initial chamber of the series was heated by exhaust steam from a steam engine, and the last vessel in the battery was connected to a vacuum pump.

Even though the Rillieux design was far more efficient and sophisticated than some of the others, there was only a slight increase in the number of multiple-effect units installed during the 1850s. While twelve units were employed in 1852, only thirteen were found on Louisiana plantations in 1859. However, in contrast to Rillieux's multiple-effect evaporators, the less complicated open steam pans and vacuum evaporators were widely employed during the 1850s— twenty-three of the former and twenty-seven of the latter could be found in Louisiana sugarhouses in 1859. In part, the reluctance to adopt the more advanced designs was due to economic considerations. A greater return on invested capital favored the adoption of far cheaper but less sophisticated apparatuses.[29] Also, the great economic advantage of the Rillieux apparatus—a 75 percent reduction in fuel consumption—was somewhat negated by the appearance of the bagasse burner during the 1850s. This furnace greatly reduced wood requirements by allowing the planter to use bagasse as a supplementary source of fuel for burners and open kettle furnaces. Furthermore, the lack of local institutions to train skilled men in the operation and repair of the multiple-effect evaporators must have

29. Bache and McCulloch, *Reports for the Secretary of Treasury,* 635–36.

been another deterrent to their widespread adoption. The existence of slave labor was another reason planters resisted these new labor-saving and energy-efficient devices.

Numerous Louisiana planters considered their slaves to be incapable of understanding and operating the sophisticated new equipment.[30] Perhaps it was for that reason that they chose the advanced-design open steam pans during the 1850s as a means of upgrading their processes while still accommodating themselves to the existing labor system. They felt that the skills acquired by the Negro slaves on the open kettles could be retained in operating the newly introduced open steam train. Indeed, planters may have even feared that upgrading the slaves' skills would lead to the questioning of authority and ultimately to insurrection. Planter Andrew Durnford, himself a free black, spurned Rillieux's attempts to install multiple-effect evaporators on his property because he did not want to lose control over his slaves.[31] Commenting about his Louisiana travels, Frederick Law Olmsted wrote in 1856:

> As commerce, or any high form of industry requires intelligence in its laborers, slaves can never be brought together in dense communities, but their intelligence will increase to a degree dangerous to those who enjoy the benefit of their labor. The slave must be kept dependent, day by day, by his master for his daily bread, or he will find, and will declare his independence, in all respects of him. This condition disqualifies the slave for any but the simplest and rudest form of labor; and every attempt to bring his labor into competition with free labor can only be successful at the hazard of insurrection.[32]

In discussions concerning the Rillieux apparatus during the 1840s, it was clear that the quality of labor necessary to operate multiple-effect evaporators was a point of contention. T. J. Packwood, one of the first owners of a Rillieux system, wrote in 1844 that the "apparatus is very easily managed, and my negroes became acquainted

30. The problems associated with a slave labor system and industrial technology are omitted in studies focusing upon work on the antebellum plantation. See Joe Gray Taylor, *Negro Slavery in Louisiana* (New York, 1969), 69–78; Robert S. Starobin, *Industrial Slavery in the Old South* (New York, 1970), 19–20, 41.

31. Ira Berlin, *Slaves Without Masters: The Free Negro in the Antebellum South* (New York, 1974), 274.

32. Olmsted, *A Journey in the Seaboard Slave States*, 591.

with it in a short time." In the 1846 *Proceedings of the Agriculturists' and Mechanics' Association of Louisiana*, the committee on the manufacture of sugar reported that the "apparatus may be worked by the hands on the plantation without any experienced sugar maker." However, a number of influential planters disagreed. Edward J. Forstall, a Louisiana planter and New Orleans merchant, wrote: "That there are some drawbacks is not to be denied. I consider it not to be at all practicable, or if so, highly imprudent to rely on slaves to work the apparatus. I think that the planter who determines to adopt the improvements should make up his mind to have in his employ at least two white persons to take charge of the apparatus during grinding season, so as to have one white person at the pans on each watch." Valcour Aime expressed an opinion similar to that of Forstall. In an 1847 issue of *DeBow's Review*, he asserted: "The Agricultural and Mechanics' Association has certainly been led into error, in stating in the report of last year that the negroes employed on a plantation can conduct these machines *without the assistance of a sugar maker*. . . . Such an error might occasion great loss to be sustained by many planters. The apparatus of Rillieux as well as that of Derosne & Cail, is too complicated to be entrusted to negroes without the active superintendence of an experienced sugar-maker."[33] It was ironic and one of the many paradoxes of the Old South that many planters argued that an evaporator invented by the black scientist Rillieux was too complicated to be operated by black men.

Several factors opposed the widespread adoption of new technology. First, the differential duty levels established in 1842, which were financially advantageous to planters employing advanced methods and equipment, were abolished in 1846. Second, for at least one group of planters the operation of relatively sophisticated apparatus, requiring careful observations and mechanical manipulations, was incompatible with their perception of the slaves' mental capabilities.

---

33. Bache and McCulloch, *Reports for the Secretary of Treasury*, 253; Agriculturists' and Mechanics' Association of Louisiana, *Proceedings of the Agriculturists' and Mechanics' Association of Louisiana: Annual State Fair, 5 January 1846* (New Orleans, 1846), 20; Edward J. Forstall, "Louisiana Sugar," *DeBow's Review*, II (1846), 344; "Valcour Aime's Results in Sugar Culture and Manufacture," *DeBow's Review*, IV (1847), 425.

But perhaps the critical factor was the immaturity of local scientific and technical institutions which could not provide a basis for sustained innovation.

Institutions and the Louisiana Sugar Industry

In his European travels, Louisiana planter Judah P. Benjamin witnessed the contributions of French institutions to the modernization of the beet-sugar industry. It was not surprising that upon his return he took steps to establish and encourage the development of local educational and scientific organizations. In an address to the Agriculturists' and Mechanics' Association of Louisiana, Benjamin asserted that the change in outlook associated with the manufacture of sugar was due to "your association. It is to your efforts that is mainly to be attributed that spirit of enquiry, which has lately aroused the agriculturists of the State from the lethargy into which for so many years they had sunk."[34] Although he spoke optimistically of the association, in truth it never reached its full potential as a strong and viable institution in the antebellum period. The organization was founded in 1842 and normally held its annual meetings in conjunction with the yearly state fair. Yet there appears to be no evidence that the association continued to meet after the late 1840s. The organization's weakness may be traced to its shortage of operating funds and its lack of local branches. In addition, it seems that the inclusion of both agriculturists and mechanics created divisions rather than the unity necessary for the emergence of a strong institution. The association, divided into two departments, agriculture and mechanics, was governed by a president assisted by numerous vice-presidents. Vice-presidents representing the sugar interests of the state were Valcour Aime, P. A. Rost, Verloin Degruy, B. M. Norman, Miles Taylor, and Maunsel White. Reporting to the association were committees on sugar, cotton, fine arts (photography), horses and cattle, mechanical inventions, and agricultural products.[35]

James DeBow thought that the organization's floundering condition could be remedied by state assistance and support. He proposed

34. "J. P. Benjamin's Address on Agriculture," *DeBow's Review*, V (1848), 57.

35. "Louisiana Agriculturists' and Mechanics' Association," *DeBow's Review*, III (1846), 181, 117; Agriculturists' and Mechanics' Association of Louisiana, *Annual State Fair, 5 January 1846*, 19–28.

that the association be permanently headquartered in Baton Rouge, "a place particularly fitted for the meeting of the Institute. It should secure a permanent Hall in the new State-house." For DeBow, "the path of the Association must be with brighter auspices; and we commend it now to the fostering care of the Legislature of our State."[36] Apparently, state sponsorship of the association never materialized, and a much-needed institution for the advancement of scientific agriculture did not develop in antebellum Louisiana.

By 1847 the Louisianians interested in the development of local scientific institutions began to shift their hopes from the Agriculturists' and Mechanics' Association to the University of Louisiana. James DeBow commented in 1847, "If we are ever to have the University of Louisiana, of which our constitution speaks, will not the planters look to it, that the institution disseminates the principles of scientific agriculture?"[37] Judah P. Benjamin, along with three others, played a key role in the formation of the University of Louisiana. In 1844 he helped prepare and present to the state convention a plan for a state university to be located in New Orleans. The plan called for the formation of a university consisting of four faculties—medicine, law, letters, and natural science.[38] After several years of delay, the plan was adopted by the legislature, and a bill creating the University of Louisiana was signed by Governor Isaac Johnson in early 1847. However, in ensuing years, the legislature proved to be unsympathetic to the financial needs of the newly established university. Each year the university faced the question of whether state legislative support would be forthcoming and whether it would be sufficient. More times than not, operating funds were inadequate. The ever-present issue of funding was undoubtedly the fundamental weakness of this institution during the antebellum period and limited its effectiveness in the agricultural community.

Although the university's fiscal status was insecure and it never succeeded in attracting large numbers of students, the University of Louisiana perceived itself as an institution responsive to local economic conditions. A University of Louisiana catalog stated "that the

36. "Southern and Western Agricultural and Mechanic Association," *DeBow's Review,* IV (1847), 422; "Louisiana Agriculturists' and Mechanics' Association," 181.

37. "Southern and Western Agricultural and Mechanics' Association," 444–45.

38. John P. Dyer, *Tulane: The Biography of a University* (New York, 1966), 21.

course and method of instruction is more directly applicable and of more ready adaptation to the pursuits and tone of society which obtains in the State of Louisiana, than at institutions among people whose domestic institutions and industrial resources differ widely from our own." In 1847 Dr. Francis L. Hawks, the first president of the university, gave a report that was subsequently published in *DeBow's Review*. Hawks asserted that science had to be taught "in its practical applications," and that agricultural and industrial chemistry, geology, civil engineering, and mechanics "are of direct and immediate interest." Commenting on the report, DeBow stated: "We admire the practical turn which is advocated, and the preference given to the *useful* over the ornamental. . . . The man is ignorant and helpless, whose learning aids him nothing in his contact with his fellows. We must learn to meet the world—to know, if not control its ways, and a proper and useful direction may be given to learning from the very moment its rudiments are gathered."[39]

The faculty of the University of Louisiana in 1850 included J. Lawrence Smith as professor of chemistry and mineralogy, John L. Riddell as professor of chemistry in the medical department, and Claudius W. Sears as professor of mathematics and natural philosophy. Smith was born near Charleston, South Carolina, in 1818 and received his preliminary education at Charleston College and the University of Virginia. After a short stint of employment with the Charleston-Cincinnati Railroad he enrolled at the medical school in Charleston, where he received a medical degree. Smith continued his education in Europe, where he received advanced training in chemistry under Jean Baptiste Dumas in Paris and Justus von Liebig in Giessen. Upon returning to the United States, Smith briefly worked for the state of South Carolina, and then, during the late 1840s, was appointed by Secretary of State James Buchanan to a post in Turkey, where he taught agricultural science and cotton culture to planters in Asia Minor. He remained at the University of Louisiana for only a short time before accepting a position at the University of Virginia in 1851.[40]

39. State University of Louisiana, *Eighth Annual Session of the Collegiate Department, 1858–59* (New Orleans, 1859), 13; "Report of Dr. Hawks," *DeBow's Review*, V (1848), 237–38.

40. *National Cyclopedia of American Biography* (1929), VI, 54.

With Smith's departure, the responsibility of teaching general chemistry at the University of Louisiana fell to John L. Riddell, another scientist with a varied professional career. Riddell was born in Massachusetts in 1807 and educated at Rensselaer Polytechnic Institute where he was awarded a bachelor's and a master's degree. In 1830 he began his scientific career as a traveling lecturer, giving lectures on chemistry, botany, and geology at such places as Ogdensburg, New York; Brockville, Kingston, and Toronto, Canada; Erie, Meadville, and Pittsburgh, Pennsylvania; and Worthington and Cincinnati, Ohio. In 1853 he was appointed to a position at the Cincinnati Medical College and a year later accepted a position at the Medical College of Louisiana, which later became a part of the University of Louisiana. In addition to this academic position, Riddell also held a federal appointment as melter and refiner at the United States Mint in New Orleans during the 1840s. He was well qualified to teach students not only theoretical chemistry, but practical techniques as well.[41]

While courses in the pure sciences were taught by Smith and Riddell, students could also earn a diploma in civil engineering under the tutelage of Claudius Sears by taking courses in calculus, natural philosophy, geology, mineralogy and chemistry, the strength of materials, building construction, topographical engineering, and drawing.[42] A later catalog revealed that under Sears's leadership the engineering course had become more firmly established in the curriculum. Freshmen and sophomores were taught drawing skills, while juniors were required to learn various subjects associated with civil engineering. In particular, students had to enroll in a course on industrial drawing "in its application to Machinery, Steam Engines, Etc," gaining skills that would be invaluable in designing sugarhouse machinery.[43] Mechanical training continued during the senior year, with discussions focusing on steam and locomotive engines.

Unfortunately, the University of Louisiana never had sufficient students or money to fully implement these engineering programs.

41. William Pitt Riddell, "Riddell Genealogy," 1852 (MS in Record Group 10, Modern Miscellany, Louisiana State Museum, New Orleans).
42. University of Louisiana, *A Catalogue for the Officers and Students of the University of Louisiana for the Academical Year 1850–51* (New Orleans, 1851), 31.
43. State University of Louisiana, *Eighth Annual Session*, 10.

On the eve of secession, the faculties of letters and natural sciences no longer existed, the result of the failure of the state legislature to support the university with enough funds.[44] The medical faculty was the only department within the university to sufficiently impress state lawmakers and thus secure adequate finances. Apparently, the pure sciences, the arts, and even engineering did not convey a strong sense of utility to those in antebellum government in Louisiana. It is clear that during the antebellum period the two institutions that could have trained scientists and engineers and promulgated French science and scientific ideology never matured enough to pursue such goals. During the 1850s technological changes in Louisiana were centered around the one local institution strong enough to influence the manufacturing community—the foundry. Sugar manufacturing, without the leadership of trained scientists and engineers, turned away from its fascination with scientific principles toward the practical labor- and fuel-saving innovations of the mechanic.

In an 1846 address to the Agriculturists' and Mechanics' Association of Louisiana, Thomas Bangs Thorpe asserted that "Agriculture finds a right hand in the mechanical arts. . . . The mechanical arts assist, and ameliorate agricultural labors, from the mighty steam engine, with its multifarious offices, through every contrivance, down to the simple gin band."[45] Until the early 1850s, the Louisiana sugar industry had depended upon northern foundries for most of the new process equipment installed in the state. The foundries of three northern cities, Cincinnati, New York, and Philadelphia, had a firm hold on the Louisiana plantation market. The Cincinnati firms of J. Nyles and Company, Joseph Goodloe and Company, and David Grifye sold a total of 281 steam engines in Louisiana alone between 1846 and 1850.[46] Philadelphia firms sold 355 mills and steam engines in Louisiana during the same period. In addition, Rillieux's multiple-effect apparatus was made and sold by Philadelphia's Merrick and Towne, "machinists of established reputation for intelligence and su-

---

44. University of Louisiana, *Annual Report, University of Louisiana, February 14, 1861* (N.p., 1861), 3.

45. Agriculturists' and Mechanics' Association of Louisiana, *Annual State Fair, 5 January 1846*, 10.

46. P. A. Champomier, *Statement of the Sugar Crop Made in Louisiana, 1849–1850* (New Orleans, 1850), 53.

perior skill."[47] In New York, the Novelty Iron Works sold steam mills and engines, Derosne evaporation apparatuses, vacuum pans, clarifiers, and granulating pans to Louisiana planters.

Maunsel White, a prominent Plaquemines Parish planter, ordered boilers, mills, and drainage pumps from New York and Pittsburgh foundries.[48] The engineers associated with these companies were a most important source of expertise in antebellum Louisiana. For example, White wrote of Adolph Aymes, an engineer employed by Knap and Totten of Pittsburgh: "My Boiler has arrived safe, but without Aymes; I can neither put it up or the Draining Machine either. . . . [I] have so much confidence in his ability that I would put it [a vacuum pan] up in my own Sugar House to try the thing, if it could be done without taking down the other [apparatus]." In return for additional services from Knap and Totten's engineers, White acted as an agent for the Pittsburgh foundry. He invited his neighbors to observe the newly installed equipment in operation in his sugarhouse, with the hope that they would order similar apparatuses.[49]

The establishment of local foundries facilitated the injection of the mechanical arts into the Louisiana sugar industry. One reason New Orleans flourished as a center of the foundry trade was the presence of large numbers of Irish and German immigrants and free blacks skilled in the manufacture of wrought iron products.[50] The foundry enabled the mechanic-inventor to practice his craft from an institutional base. J. A. Leon, a perceptive observer of the Louisiana sugar industry, wrote in 1848 that local foundries were changing from workshops for the repair of steam-powered mills and process apparatuses to establishments manufacturing new equipment. He wrote, "Opposite the city of New Orleans, in a place called Algiers, on the right side of the Mississippi river, another extensive estab-

47. Bache and McCulloch, *Reports for the Secretary of Treasury*, 252.

48. Maunsel White to Alfred Stillman, May 15, 1845, White to Messrs. Stillman Allan Co., July 21, 1845, White to Messrs. Stillman Allan Co., November 17, 1845, White to Adolph Aymes, May 27, 1847, White to Messrs. Stillman Allan Co., November 22, 1847, White to Messrs. Knap and Totten, September 24, 1849, all in Maunsel White Papers, II (1845–50), Southern Historical Collection of the University of North Carolina Library at Chapel Hill, hereinafter cited as SHC.

49. Maunsel White to Messrs. Knap and Totten, July 2, 1847, and May 27, 1847, in White Papers.

50. Christian, *Negro Ironworkers in Louisiana*, 47–61.

lishment for the construction of machinery is nearly finished; some Louisiana parishes are also building new workshops for the reparing [sic] of the machinery."[51] Leon was describing Belleville Iron Works. By 1850 two other local companies were manufacturing new apparatuses: the Phoenix Foundry, of Gretna, Louisiana, and Leeds and Company of New Orleans.

A number of other local foundries began to have an impact upon the Louisiana sugar industry. Advertisements for the manufacture of sugar-making apparatuses first appeared in Champomier's *Statement of the Sugar Crop Made in Louisiana* in the mid-1850s. Daniel Edwards, located in New Orleans, had had success in selling open steam trains to a number of planters. Edwards, a manufacturer "of every Description of Copper, Tin, Sheet Iron and Brass work," was "now ready to contract for the making of steam trains, clarifying and evaporating pans, filters, juice boxes and everything appertaining to the sugarhouse."[52] And the Ferdinand W. C. Cook Company, located at the foot of Canal Street in New Orleans, advertised and sold a line of mechanical equipment to sugar planters of the region, including six-foot-diameter portable boilers, steam engines, and rotary pumps capable of discharging from seven hundred to one thousand gallons of water per minute for the drainage of cane fields. This heavy machinery would be typically delivered by steamboat to the planters' docks.[53]

It appears that the new local foundries associated with the sugar industry were of two basic types. One manufactured equipment primarily from patent rights assigned to it; the other, in addition to making all types of steam-related apparatuses, fabricated equipment invented by an employee or the owner. The Belleville Iron Works fits into the first category. One of its products was a bagasse furnace patented by the New Yorker Alfred Stillman.[54] The second type of foundry may be represented by the Boston Steam Engine Company,

51. Leon, *On Sugar Cultivation*, 71.

52. P. A. Champomier, *Statement of the Sugar Crop Made in Louisiana, 1854–1855* (New Orleans, 1855), back cover.

53. "Advertisement Letter, W. C. Cook" [1859?], Robert R. Barrow Family Papers, Manuscripts Department, Special Collections Division, Howard-Tilton Library, Tulane University, New Orleans, La.

54. *Stillman's Patent Bagasse Furnace* (New Orleans, 1855).

which offered the services of inventor and mechanic Samuel H. Gilman. The company made equipment based on Gilman's patents for steam trains and bagasse furnaces. Furthermore, it either constructed or carried in stock "sugar mills and steam engines of all sizes, Every Description of Steam Apparatus, Steam Boilers of all kinds," and other equipment.[55]

During the 1850s Gilman designed two types of open steam apparatuses and a bagasse furnace. He based his apparatus not only on his observations gained from practical experience, but also on simple mathematical calculations for the determination of relative efficiency.[56] His steam train designs attempted to rectify the common problems encountered by the sugar planter, including inefficient heating by the steam coils, inaccessibility of the coils for cleaning, and condensate formation in these coils. By a clever positioning of steam pipes, discharge tubes, and false bottoms, Gilman's apparatus was an answer to frequent complaints. His work was similar to the efforts of a number of other local inventors employed in foundries. H. O. Ames, whose offices were located in New Orleans, patented in 1859 a "Star Pan," or steam train designed to obviate the same problems that had inspired Gilman's design.[57]

The decade of the 1850s marked the rise of the local foundry and the widespread adoption of its products—particularly open steam trains—on the Louisiana sugar plantations. After the decade of the 1840s, in which chemical knowledge and the significance of science and sophisticated mechanical apparatuses were apparently accepted, there was a return to practical designs based on experiential knowledge. This was in part because the institutions most crucial to the continued injection of science into Louisiana were conspicuously weak. However, a center for the application of practical knowledge, the foundry, emerged as a strong institution during the 1850s with an increased significance to the local sugar industry. Individuals continued to design fuel- and labor-saving devices, but this activity was centered around the institution of the practical man—the foundry. Open steam trains and vacuum pans, rather than the more compli-

55. Gilman, *Begass Considered as Fuel for Making Sugar*, 21.
56. *Ibid.*, 9–7.
57. Ames, *H. O. Ames' Improved Method*, 2–12.

cated multiple-effect evaporators, were installed in increasing numbers on Louisiana plantations by local mechanics. These designs were the product of the mechanic's observations, coupled with trial-and-error experiments. The popularity of this equipment was in part due to its relative low cost, its simplicity of operation, and its compatibility with planters' views concerning the capabilities of their Negro slaves.

Why did scientific, technical, and business institutions related to the Louisiana sugar industry fail to mature during the period prior to the Civil War? Part of the answer lies in an underdeveloped transportation system in the region that kept sugar planters southwest of New Orleans in relative isolation. And the fact that the cultivation and manufacture of sugar depended on slave labor placed restraints on technological development. Also, between 1830 and 1861 the state of Louisiana and the federal government did little to encourage the development of institutions such as agricultural experiment stations or colleges for training students in the mechanical arts or agriculture. Perhaps the major factor inhibiting the creation of allied institutions was an economic one: as long as prices were high and competition virtually nonexistent, no edge in the marketplace was necessary.

The Civil War marked a watershed for the Louisiana industry by precipitating tremendous changes in ownership and in the traditional labor system. Furthermore, concurrent with the Civil War, a new international sugar market was rapidly developing. The market, centered in Europe, was dependent on beets rather than cane as a source of sugar. Possessing generous government subsidies and efficient process machinery and production methods, the beet-sugar manufacturers of the 1860s were a serious threat to a Louisiana sugar industry that had been reluctant to change its organization and technology during the 1840s and 1850s and had been subsequently disrupted by the war. It would be this new and extremely complex business world that gradually forced the independent planter to organize in order to promote his economic interests and ensure financial survival.

3

Post–Civil War Developments and the Rise of the German Beet-Sugar Industry

The Civil War marked the end of a period of steady growth for the Louisiana sugar industry.[1] The disruption of cultivation and manufacturing activities by changes in local government and military operations proved devastating. Sugar production, which had peaked at 264,000 tons in 1861, fell to 9,950 tons in 1865. In the period immediately after the Civil War, the Louisiana planter was faced with formidable difficulties and challenges. The certainties associated with slavery were now replaced by a new, untried free-labor system. Many of the best antebellum plantations were abandoned and in disrepair, and the unsettled political and social conditions cast a shadow over those who planned to rebuild sugar estates.[2] Sugar planters required large sums of capital and an abundant supply of labor, both of which proved to be scarce throughout the Reconstruction. Lands needed to be cleared and ditched, livestock replaced, and new agricultural implements and machinery purchased. The large number of foreclosures on sugar plantations was not the direct consequence of the Civil War; however, three years of floods and bad

1. Charles P. Roland, "Difficulties of Civil War Sugar Planting in Louisiana," *Louisiana Historical Quarterly*, XXXVIII (1955), 40–62.

2. Sitterson, *Sugar Country*, 231–42. Political histories of the period include: Joe Gray Taylor, *Louisiana Reconstructed* (Baton Rouge, 1974); Ella Lohn, *Reconstruction in Louisiana after 1868* (New York, 1930).

weather after 1865 forced banks like the Citizens National Bank of New Orleans to reluctantly call in loans.[3] As a result of these hard times, a large number of northern carpetbaggers with outside capital purchased property in the region and applied their commercial talents to the sugar business. For example, New York City coffee and tea merchant John Dymond began spending large amounts of money in 1869 to purchase new evaporation kettles, clarification equipment, and agricultural implements and to pay laborers to ditch his fields in Plaquemines Parish.[4] In a sense, what changed dramatically after the war was not so much the background of the planters, for many of the antebellum plantation owners were originally from the northern states, but rather the international sugar market. And while France served as the focal point of new science and technology in this industry before the war, after 1865 Germany emerged as the center of scientific research and new process development.

The term *modernization* can mean many things; however, in this study modernization means the various steps that were taken by individuals associated with the Louisiana sugar industry to keep it competitive in world markets. Political, organizational, scientific, and technological components figured in the development of the late nineteenth-century Louisiana sugar industry. To a large degree the modernization of the Louisiana sugar industry between 1865 and 1900 was the consequence of the injection of German ideas into this gradually emerging large-scale enterprise of the Deep South. German scientists and engineers provided the technical leadership that not only created a new European industry, but also later formed the basis for future developments in post-Reconstruction Louisiana. Thus to fully understand modernization in Louisiana, the beet-sugar industry in Germany must be examined first.

Science, Technology, and the Rise of the German Beet-Sugar Industry

In 1747 the German chemist Andreas Sigismund Marggraf (1709–1782) published a study entitled *Expériences chymiques faites dans le dessein de tires un veritable sucre des diverses plantes qui, crois-*

3. Personal communication with Glenn R. Conrad.
4. "Lists of Debits and Credits" [1869?] (MS in Dymond Family Papers, Folder 252, Historic New Orleans Collection, hereinafter cited as HNOC).

sent dans notres contrées, which described methods to extract sugar from various varieties of beets. Franz Achard, Marggraf's successor at the University of Berlin, employed his predecessor's early investigations to develop a thriving beet-sugar industry during the Napoleonic Wars. However, these small factories collapsed after Waterloo when Britain poured her stored supplies of colonial sugar into the Hanse Towns.[5] As a result of depressed grain prices during the 1830s, the German landed gentry began to reapply Achard's methods and to cultivate root crops. Since a twofold profit was possible by first harvesting the beets and then refining the product, large estate owners were particularly attracted to the possibilities offered by the sugar beet. They obtained the capital for these ventures from merchants located in the larger cities of Saxony and Prussia. The resultant boom in the German beet-sugar industry was also the consequence of the use of improved farm machinery for the deep cultivation and drilling necessary for planting.[6] The nineteenth-century German beet-sugar industry was unique—heavily dependent on production advantages gained from technology, yet intimately connected to a traditionally conservative agricultural system.

In many respects, the various operations associated with the manufacture of beet sugar were quite different from those used to process cane sugar. First, the beets were washed and cleaned in a machine consisting of revolving drums. The washed beets were then either sliced or ground to a pulp, depending on whether juice extraction was accomplished by pressing, centrifugal force, maceration, or diffusion. Once the juice was separated from the roots, it was purified by the addition of regulated quantities of lime (saturation), followed by successive additions of carbon dioxide (carbonatation). In order to separate the raw sugar and syrup, the resulting viscous solution was spun in large centrifuges. The supernatant syrup could be further boiled under vacuum to yield additional raw sugar and molasses.[7]

Between 1860 and 1880 improvements in the manufacture of sugar

5. J. H. Clapham, *The Economic Development of France and Germany, 1815–1914* (Cambridge, 1921), 87.

6. Theodor Schuchart, *Die Volkswirtschaftliche Bedeutung der Technischen Entwicklung der Deutschen Zuckerindustrie* (Leipzig, 1908), 8; Clapham, *Economic Development of France and Germany,* 51.

7. Rudolf Wagner, *A Handbook of Chemical Technology,* trans. William Crookes (New York, 1892), 685–708.

occurred in almost every area of the plant. New types of blades were designed for the slicers, advanced high-pressure filter presses separated small amounts of sugar from sediments, and bone black was replaced by gravel filtrations.[8] The technology of evaporation was enhanced by the efforts of Norbert Rillieux, now in Paris, and Julius Robert. Through their efforts, steam technology was employed at several key points in production—during the operations of diffusion, saturation, and filtration.

The introduction of the diffusion process by Julius Robert in 1865 was a crucial application of innovative technology to the beet-sugar industry. The new diffusion process, a result of Robert's twenty years of investigations, was quickly accepted by German manufacturers. The old method of obtaining sugar juice began with shredding the beets, placing the pulp in linen cloths, and applying pressure by means of either hydraulic, roller, or filter presses. In the diffusion process, beet roots were cut into one-millimeter-thick slices, called cossettes, which were then digested at elevated temperatures in closed iron cylinders. These vessels were arranged in groups of ten or twelve, called batteries. Once a cylinder was charged with beet cuttings, it was filled with a warm solution of sugar juice. The sugar was transferred from the beets to the aqueous phase, and the extracted sugar was circulated to a second diffuser, where it was exposed to fresh cuttings. The operation was repeated until the juice passed through the entire battery. The process operated in a semicountercurrent fashion: clean water entered the cell containing the nearly exhausted cossettes and traveled through each cell in succession.

The beet-sugar industry was revolutionized not only by the introduction of new process apparatuses, but also by the application of chemical science.[9] Between 1861 and 1877 quantitative methods replaced rule-of-thumb practices in the sugarhouse. Specific gravities, acid-base titrations, and polarimetry supplanted qualitative schemes based on the sense observations of odor, color, and the appearance of a precipitate. Agricultural problems were systematically explored, and large-scale beet processing was often modified after examining data obtained from controlled process runs.

8. "Über Neuerungen in der Zuckerfabrikation," *Dingler's Polytechnisches Journal,* CCXXXIV (1879), 407.

9. Schuchart, *Die Volkswirtschaftliche Bedeutung der Technischen,* 66–67.

The German chemist Justus von Liebig wrote that methodology was a key to the beet-sugar success and that cane-sugar producers had numerous advantages that were wasted because of their failure to adopt new technology. Liebig stated: "The sugar beet manufacturers have advantages over the colonial planters because they use better methods; that is, they use less labor . . . and maybe they have greater knowledge."[10] Only during the last quarter of the nineteenth century were the scientific and technical advantages, closely associated with the European beet-sugar industry, gradually assimilated into the Louisiana sugar business.

In Germany the sugar industry developed a means of plant control by the early 1860s, as evidenced by the 1863 book of tables by O. Frese, entitled *Beiträge zur Zuckerfabrikation*. In the introduction, Frese warned: "The sugar house manufacturer cannot work according to his old, careless ways, without having to complain about losses. He must be certain that through analysis and calculations he will obtain a profit and avoid liabilities."[11]

Frese's book was divided into three sections. The first section contained a series of tables that correlated the concentration of sugar in solution with specific gravity. A second group of tables dealt with the use of the polarimeter. Frese's tables related sucrose content in product sugar, molasses, and by-products with the measured angles of rotation. This rapid, specific, and quantitative method was later widely employed, including in Louisiana sugarhouses, and facilitated process control. The last section of Frese's treatise consisted of a number of tables necessary for the analytical control of several specific operations in the plant. When making a difficult decision about the addition of chemicals, these tables were a definite aid to the sugarhouse operators. For example, the knowledge of the specific gravity of a sugar juice would lead to the determination of the proper lime addition. The charts also provided information on how much sugar could be pressed from different grades of beets. Furthermore, the tables correlated the amounts of CO_2 available from a certain weight of $CaCO_3$, and the amount of $CaCO_3$ contained in different types of bone black.

10. Justus Liebig, *Chemische Briefe* (Leipzig, 1878), 100.
11. Frese, *Beiträge zur Zuckerfabrikation . . . Ein Hülfsbuch für Fabrikherren, Direktoren und Siedemeister* (Brunswick, 1865), v.

The polarimeter was used to determine the percentage of sugar in beets, beet residue after diffusion, extracted juices, and waste water, making it an important tool for process modifications. The polariscope enabled the manufacturer to run a large number of samples reliably and at low cost, making the instrument well suited for monitoring changes in a large-scale process. Data could be reported quickly, and sugar content could be determined in a variety of different samples. In the 1860s numerous investigations were undertaken in an effort to improve processes for refining molasses. Typically, these investigations included a large number of reaction parameters. For example, industrial scientists systematically studied the effects of different temperatures and flows.[12] In an attempt to avoid the use of expensive bone black, polarimetric analysis was used during the late 1870s and early 1880s to evaluate alternate means of filtration.[13]

Specific gravity measurements were also employed extensively during the development of the diffusion process. During the mid-1860s specific gravity data was a criterion for the proper use of the various valves, and these measurements determined the flow patterns among the diffusion cylinders. Temperature regulation of the diffusion process was determined during production runs, and chemists relied upon specific gravity measurements for the establishment of optimum operating conditions.[14]

The introduction of scientific methodology played a key role in the success of the German beet-sugar industry. Developments of new designs were often influenced by data provided by analytical chemistry. Chemical analysis transformed the central sugar factory by rapidly supplying answers to problems that had previously perplexed the operator. Before the 1860s it appears that the beet-sugar plant was run by experience; after 1860 management increasingly relied on analytical data to control processes. And the significance of this revolution in Germany is that it was the reliance upon exact analytical data in process monitoring and in process development

12. Karl Stammer, *Jahres-Bericht uber die Untersuchungen und Fortschritte auf dem Gesamtgebiet der Zucker Fabrikation* (Breslau, 1868), 303–11.

13. "Über Neuerungen in der Zuckerfabrikation," *Dingler's Polytechnisches Journal*, CCXLII (1881), 208–11.

14. Karl Stammer, *Jahres-Bericht über die Untersuchungen und Fortschritte auf dem Gesamtgebiet der Zucker Fabrikation* (Breslau, 1875), 5–6.

that helped a modern sugar industry to emerge in Louisiana between 1880 and 1900. Exact weights and volumes were now the crucial parameters of plant efficiency and product quality.

What was the chief driving force behind this revolution in beet-sugar methods and manufacturing techniques? Bounty legislation (first established in Germany in 1861 and later adopted by other European nations) proved to be a primary stimulus for the growth of the European beet-sugar industry. Bounties were government subsidies paid to the producer in addition to the price received from consumers in the marketplace. This rebate was paid on exported sugar, and depending upon the tax rates, a continental producer could sell his sugar abroad at prices below production cost while still realizing a profit. In effect, bounties created low prices for consumers in nations having low import tariff rates (such as nineteenth-century Britain) and higher market prices for the exporting nations.

As a result of the sugar bounty, Germany, Austria-Hungary, France, and Prussia became competitors in the manufacture of beet sugar. In 1840 tropical cane sugar had dominated world trade, but in large part because of financial incentives, by 1870 the beet-sugar industry accounted for one-third of all sugar produced (Tables 1 and 2).[15] Thus, during the decade of the 1860s—one of great trouble for the Louisiana industry—the market share of European beet sugar expanded dramatically. Germany nearly doubled its output, France nearly tripled its production, and Russia increased its sugar tonnage approximately fivefold. While new technology played an important role in the growth of the sugar industry in these countries, bounties and high tariffs—the reflection of an economic nationalism advocated by Louis Napoleon, Bismarck, and Alexander II—created an economic environment that stimulated agriculturists, industrialists, and financiers.

To complicate matters for the planter in post–Civil War Louisiana, he was faced with competition not only from European beet-sugar producers, but also from tropical cane-sugar manufacturers. While the British and French sugar-producing colonies generally had little impact upon the American economy, Cuba became a primary

15. Alfred Chapman and Valentine Waltran Chapman, "Sugar," *Encyclopaedia Britannica* (1911 ed.), XXVI, 32–48; Deerr, *History of Sugar*, II, 492–94.

Table 1

THE WORLD'S TRADE IN CANE AND BEET SUGAR
(thousands of tons avoirdupois)

Year	Cane	Beet	Total	% Beet
1840	1,100	50	1,150	4.35
1850	1,200	200	1,400	14.29
1860	1,510	389	1,899	20.48
1870	1,585	831	2,416	34.40
1871–72	1,599	1,020	2,619	38.95
1872–73	1,793	1,210	3,003	40.29
1873–74	1,840	1,288	3,128	41.17
1874–75	1,712	1,219	2,931	41.59
1875–76	1,590	1,343	2,933	45.78
1876–77	1,673	1,045	2,718	38.44
1877–78	1,825	1,419	3,244	43.74
1878–79	2,010	1,571	3,581	43.87
1879–80	1,852	1,402	3,254	43.08

Table 2

EUROPEAN BEET SUGAR PRODUCTION
(metric tons)

Year	Germany	France	Austria-Hungary	Russia
1840	14,206	29,939	—	—
1850	53,349	76,151	—	13,100
1860	126,526	100,876	77,000	26,200
1870	262,987	282,109	213,000	125,000
1875	346,196	462,300	321,000	178,000
1880	594,360	331,000	533,000	304,000

supplier of sugar to the United States during the Reconstruction and post-Reconstruction periods (Table 3).[16]

During the 1870s imported cane sugar from Cuba and beet sugar from Europe were usually processed by east coast refineries in the United States. Concurrently, Hawaii became a primary supplier to United States refineries on the west coast. With the passage of a reciprocal trade agreement between the United States and Hawaii in 1876, certain commodities, including sugar, were admitted into the

16. Deerr, *History of Sugar*, I, 131.

Table 3

CUBAN CANE SUGAR PRODUCTION, 1840–1880

Year	Metric tons
1840	160,891
1850	223,145
1860	447,000
1870	726,000
1875	718,000
1880	530,000

Table 4

HAWAIIAN ISLAND SUGAR PRODUCTION, 1870–1882

Year	Metric tons
1870	8,385
1871	9,720
1872	7,690
1873	10,300
1874	10,970
1875	11,200
1876	11,640
1877	11,140
1878	17,240
1879	21,870
1880	28,400
1881	41,800
1882	51,000

United States duty free. Once the American market was opened to Hawaiian planters, Hawaii's sugar industry expanded dramatically; output more than quadrupled between 1876 and 1882 (Table 4).[17]

In addition to this transformation in the international sugar trade, changes were also taking place in the patterns of sugar use. Per capita sugar consumption in the United States increased from seventeen pounds per person during the 1840s to an average of thirty-six pounds per person in the 1870s (Table 5).[18] The rise in the demand for

17. Lippert S. Ellis, *The Tariff on Sugar* (Freeport, Ill., 1933), 50; Deere, *History of Sugar*, I, 131.
18. U.S. Department of Agriculture, *Report of the Commissioner of Agriculture for the Year 1878* (Washington, 1878), 63–65.

Table 5

PER CAPITA U.S. SUGAR CONSUMPTION

Decade	Pounds per person
1790–1799	6.64
1800–1809	10.06
1810–1819	8.03
1820–1829	9.28
1830–1839	12.89
1840–1849	17.08
1850–1859	29.00
1860–1869	26.96
1870–1879	35.64

Table 6

CONSUMPTION OF SUGAR IN THE U.S.
(metric tons)

Year	Sugar consumed	Sugar imported	Percent imported
1860	415,281	296,250	71.3
1861	363,819	241,420	66.4
1862	432,411	241,411	55.8
1863	284,308	231,398	81.4
1864	220,660	192,660	87.3
1865	350,809	345,809	98.6
1866	391,678	383,178	97.8
1867	400,568	378,068	94.4
1868	469,533	446,533	95.1
1869	492,899	447,899	90.9
1870	530,692	483,892	91.2
1871	633,314	553,714	87.4
1872	637,373	567,573	89.0
1873	652,025	592,725	90.9
1874	710,369	661,869	93.2
1875	685,352	621,852	90.7
1876	638,369	561,369	87.9

sugar closely paralleled increased consumption of coffee and tea, as well as the emergence of the rapidly growing processed food industry, which included confectioneries, hard candies, jams, and jellies. With its population undergoing a dietary revolution incorporating increasing amounts of sugar, the United States relied on foreign sugars to meet consumer demand (Table 6). Prior to the Civil War roughly 70 percent of all sugar consumed in the United States came from abroad; this level rose to approximately 90 percent by the early 1870s. Louisiana could hardly claim to be the "sugar bowl" of the nation. Indeed, its once lucrative trade was now threatened by a rapidly emerging European giant.[19]

The Diffusion Process in Louisiana, 1869–1877

Louisiana planters were not oblivious to the changing international scene. Alexis Ferry, a St. James Parish resident and son-in-law of Valcour Aime, shifted his thoughts from innovations in the design and layout of open kettles that he had designed before the war to advocating the application of beet-sugar technology to cane-sugar processing.[20] Louisiana planters were first exposed to the Robert diffusion process in 1869 when the United States consul in Vienna, a Dr. Canisius, arrived in New Orleans to convince planters to adopt Julius Robert's process on their plantations. Canisius had been given Robert's patent powers and carried with him the three-year report describing the benefits of the process both in extraction efficiency and product purity that were obtained at the Aska Sugar Company of Madras.[21] However, Canisius encountered indifference during his tour of the sugar parishes, and his efforts ended in failure. In 1871 a similar mission was conducted by a Mr. Wessely from New York. Planters, faced with severe shortages of both labor and capital, were disinterested in the innovation, and Wessely also failed.[22]

The Robert process was again introduced in Louisiana in 1873 when Rudolph Sieg, a German immigrant residing in New Orleans, persuaded a few European and American friends to raise the capital necessary for the purchase of a diffusion apparatus and its installation on James P. Kock's plantation, Belle Alliance, in Ascension Par-

19. U.S. Department of Agriculture, *Report of the Commissioner of Agriculture for the Year 1877* (Washington, 1877), 24.

20. Journal of Alexis Ferry, December 3, 1869, and January 28, 1873 (MS in Manuscripts Department, Special Collections Division, Howard-Tilton Library, Tulane University).

21. A summary of the three-year report of the Aska Company written by Ferdinand Kohn appears in "The Diffusion Process," *Sugar Cane*, II (1870), 270–74.

22. The early efforts of individuals attempting to introduce the diffusion process into Louisiana is summarized in Otto Kratz, *The Robert Diffusion Process Applied to Sugar Cane in Louisiana in the Years 1873 and 1874* (New Orleans, 1875).

ish. In the late fall of 1873, apparatus and engineers arrived in New Orleans. Ten vessels, eight feet high and six feet in diameter, had been transported by land across Europe to Bremerhaven, shipped by steamer to New Orleans and by steamboat up the Mississippi, finally arriving by raft at their destination on Bayou Lafourche.[23] On December 18, 1872, the apparatus was finally ready, but in the meantime, the cane fields had been frostbitten twice. The extracted juice passed from the clarifiers to five open kettles and an inefficient vacuum pan. In the experiment that followed, four hundred tons of cane were processed by the diffusion method. An extraction efficiency of almost 85 percent was obtained—outstanding results when compared to the 66 percent efficiency obtainable from the most efficient mills of the time.

In spite of this apparent success, the first trial was plagued with mechanical difficulties, and the savings in labor were not as great as anticipated. Apparently the design had been based on that of the equipment used at Aska, where labor cost five cents per day and labor-saving devices were of little value. The apparatus openings were too small: it took four to six men nearly three-quarters of an hour to empty one cell, and it could not be refilled in less than twenty-five minutes.[24] Yet apparently the process had demonstrated sufficient promise; in March, 1874, Rudolph Sieg, Dr. Otto Kratz, and Robert formed the Julius Robert Diffusion Process Company. The new corporation, capitalized at $100,000, had a host of officers and stockholders that were no doubt part of the dynamic German community of New Orleans during the post–Civil War era.[25] The company contracted with Archibald Mitchell of the Leeds and Company foundry for an improved apparatus. It was intended that one battery would be placed at Kock's Belle Alliance plantation, while another would be installed at E. C. Palmer's Southwood.[26]

Mitchell improved the apparatus in several ways. To begin with, he

23. *Ibid.*, 10.

24. J. B. Wilkinson, Jr., *Wilkinson's Report on Diffusion and Mill Work in the Louisiana Sugar Harvest of 1889–90* (New Orleans, 1890), 34–36. See also *LP*, X (1893), 83–84.

25. John Frederick Nau's *The German People of New Orleans, 1850–1900* (Leiden, 1958) does not mention the diffusion venture or any of the participants.

26. Wilkinson, *Wilkinson's Report on Diffusion and Mill Work*, 34–36.

designed a movable cane carrier that discharged "its load at various points of delivery at the discretion of the operator."[27] This equipment enabled the workman to direct the sliced cane to any of the diffusion vessels. Mitchell also redesigned the diffusion cells. By altering the lower section of the diffusion cell, he made it easier to unload the exhausted cane slices. Under the new system, three men could fill a cell in 13½ minutes and discharge it in 4 to 6 minutes. Kratz observed: "From the very start, the whole machinery worked like clockwork, and it is to be noticed as a fact well worth recording, that during this whole week the sugar process never had to wait one moment for the new machinery. It was a triumph of engineering skill, almost unprecedented."[28]

However, Kratz's optimistic comments overlooked several serious problems that were experienced at Belle Alliance and Southwood. Although the diffusion apparatus had a 40 percent greater efficiency over millwork, the lack of sufficient evaporation capacity resulted in Kock's continued preference for milling. And at Southwood a chemical problem surfaced for the first time. Apparently the product syrup did not crystallize in the open kettles. In the first run about ten hogsheads of sugar were made, but subsequent crystallization problems became so acute that it was concluded that the diffusion process could not be employed with either open kettles or open steam trains. In 1875 both devices were removed to the Louisa plantation, where they were placed in parallel and subsequently connected to double-effect evaporators. At Louisa, the Diffusion Process Company contracted to purchase cane along the river. However, transport problems, including the sinking of one barge, resulted in a delay of ten days between cutting and processing.[29] The future of the company was in peril, since it lacked the cash to meet the demands of creditors. Completely failing to produce sugar during the first runs, the company was bankrupt. However, during a fourth trial the yield of sugar per ton of cane exceeded that of any previous mill or diffu-

27. U.S. Patent Office, *Specifications of Patents* (Washington, 1876), 208–209. "Movable Sugar Carriers," February 1, 1876, U.S. Patent 173,038. "Diffusion Vessels for Extracting Cane Juice," February 1, 1876, U.S. Patent 173,039.

28. Kratz, *The Robert Diffusion Process*, 36.

29. *Louisiana Sugar Bowl* (New Iberia, La.), July 28, 1879.

sion trial in Louisiana, renewing the hope of a few planters that the process could be commercially feasible.

Between 1877 and 1879 a number of inexperienced planters tried the machinery without success; at the end of 1879 the apparatus was sold. Although the first chapter in the history of diffusion in Louisiana ended in failure, the benefits of this controversial process continued to be discussed. After 1878 a number of individuals felt that the chemical and mechanical deficiencies of the process could be overcome. Henry Studniczka, a recent immigrant and graduate of the Vienna Polytechnic Institute, noted that improper crystallization "had formerly happened in Europe; and that it can be avoided and so easily explained, . . . refer to any good book on chemistry for further information."[30] While a number of other planters argued along lines similar to that of Studniczka, or pointed to the success of diffusion at Aska or in Europe, the process still had a number of formidable critics.

Agricultural Chemistry in Louisiana During the Reconstruction Period

Just as the beet-sugar industry in Europe served as a model for proposed process changes in the Louisiana sugar industry, Louisiana's agricultural activities were also influenced by European ideas in science and technology. In particular, sugar planters noted the apparent success of European agricultural scientists in increasing the sugar content in beets by the application of chemical fertilizers. In the antebellum period planters employed plowed-under peavines, properly composted stable manure, cane tops, corn husks, bagasse, and the skimmings of sugar kettles as fertilizers. Interest in improved agricultural practices at this time was not isolated to the Louisiana sugar bowl. Throughout the South, farmers and planters were intrigued by the potential of chemical fertilizers to restore their worn-out soils and increase crop yields. The most influential scientific agriculturist during the Reconstruction was Georgian David Dickson, whose ideas on the subject were communicated in the *Southern Cultivator* and his *Practical Treatise on Agriculture* (1870).[31] By the early 1870s,

---

30. *Ibid.*, April 15, 1878.
31. Chester McArthur Destler, "David Dickson's 'System of Farming' and the Ag-

a number of recently established New Orleans companies marketed bone dust, cottonseed meal, and guano fertilizers. However, planters were usually confused as to when and how to apply these materials, as well as being unsure of their actual cost benefits. The editor of the *Louisiana Sugar Bowl* stated in 1874 that "while planters are actually groping in the dark, yet they are thirsting for information on this all important question. No doubt the yield of sugar can be increased, . . . but *more light* on the subject must be given first, before fertilizers are generally used."[32]

By the mid-1870s, agricultural scientists propounded several theories concerning the action of fertilizers, and these ideas gained some acceptance among Louisiana sugar planters. One popular theory was based on the so-called mineral system.[33] Simply stated, the constituent elements of a plant reflected its nutritive needs. Therefore, the analysis of a plant's ashes revealed the relative proportions of substances necessary for its proper growth. Most Louisiana planters remained skeptical of this theory, particularly because they felt that adherence to it would result in the cultivation of cane not only high in saccharine content, but also rich in minerals that would cause subsequent extraction difficulties. Another fertilizer theory, attractive to many planters, was the "complete manure theory" of agricultural scientist George Ville of Vincennes, France. The influential publisher Louis Bouchereau stated:

> If we are to be guided by the opinion of scientists and reports of practical commissioners, appointed by European governments to make experiments in this regard, we find that the preparation of M. George Ville stands at the head of the list, and its component parts are phosphate of lime, nitrate of potash and sulphate of lime, in due proportions. . . . The quantity per acre should be from 600 to 800 pounds, according to the judgement of the planter, who is supposed to be acquainted with the nature, needs and richness of his own soil.[34]

---

ricultural Revolution in the Deep South, 1850–1885," *Agricultural History*, XXXI (1957), 30–39.

32. *Louisiana Sugar Bowl* (New Iberia, La.), May 7, 1874.

33. *Ibid.*, July 3, 1873, and June 4, 1874.

34. L. Bouchereau, *Statement of the Sugar and Rice Crops Made in Louisiana in 1875–76* (New Orleans, 1876), XVII.

Since he felt that chemical methods had inherent limitations, Ville refined his ideas through experimental field trials.[35] Ville claimed that while plants needed many different nutrients, four were crucial to proper growth: nitrogenized matter, phosphate of lime, potash, and lime. The collective action of these four substances in their proper proportions, free from needless materials that hindered their uptake by the plant, constituted Ville's "complete manure." The manure's effectiveness was due to the collective action of the four associated compounds. However, for each particular plant, one substance of the four dominated plant activity, and agricultural practice determined the minimum proportion of subordinate compounds and the optimum dosage of the dominant substance. Ville determined that optimum beet production could be obtained by raising the kilograms of nitrogen per hectare, while holding constant the quantities of phosphate of lime, potash, and lime.

Between 1875 and 1880 a number of Louisiana planters used Ville's basic techniques to investigate the proper proportions necessary for the growth of the saccharine-rich cane. One study was conducted by Henry Studniczka.[36] To the surprise of some, Studniczka found that the most commonly applied fertilizer in the cane field, cottonseed meal, gave the worst results in his trials; soluble bone meal resulted in the highest percentage of saccharine in cane. Further, the application of Ville's complete manure formula led to better-than-average saccharine content.

T. Mann Cage, a Terrebonne Parish planter, adhered to Ville's theories and maintained an active correspondence with the French scientist. In 1876 Cage began to study the effects of fertilizers on cane by using carefully controlled test plats. He used several analytical chemical techniques, including polariscopic analysis, to determine the proper proportion of fertilizers necessary for a high saccharine content in cane.[37] By 1878 Cage had concluded from his experiments

35. George Ville, *The School of Chemical Manures: Or Elementary Principles in the Use of Fertilizing Agents* (Philadelphia, 1872), 49. See also George Ville, *Chemical Manures: Agricultural Lectures Delivered at the Experimental Farm at Vincennes, in the Year 1867*, trans. F. J. Howard (Atlanta, 1871).

36. *Louisiana Sugar Bowl* (New Iberia, La.), December 27, 1877.

37. Louis J. Bright, *New Orleans Price Current: Yearly Report of the Sugar and Rice Crops of Louisiana. Crop Year 1876–77* (New Orleans, 1877), XXVII.

that a composite of superphosphate of lime and the ashes of cotton-seed hulls was the most economical form of fertilizer containing soluble ammonia, potash, and phosphoric acid.

Between 1876 and 1878, Daniel Thompson, a planter in St. Mary Parish, employed C. A. Goessman of the Massachusetts Agricultural College to study fertilizer requirements. Goessman, a former pupil of the German chemist Friedrich Wöhler, held a doctorate from Göttingen and had previous experience in the cane-sugar industry. In 1865 Goessman had studied both the soil condition and methods of cane-sugar manufacture in Cuba.[38] In his experiments for Thompson, Goessman examined the effects of animal matter, cottonseed, ammonium sulfate, potassium nitrate, and acid phosphate of lime on eighteen one-acre plats, eleven containing new cane and seven containing ratoons (sprouting cane stubble). Goessman's resulting data included quantity of fertilizer applied, composition of its soluble nutrients, total cost of the fertilizers, weight of cane produced, gallons of juice produced, juice density and polarization, percent saccharine content, and total weight of sugar obtainable. The first study concluded that planters should fertilize with cottonseed meal at the time of first plowing and follow with additions of ammonium sulfate, lime, and potash after the cane had sprouted.[39]

Overall, the studies conducted at the Thompson and Cage plantations were inconclusive and led to additional controversy. One critic felt that the determination of saccharine content by polarimeter (rather than by carefully determining the amount of extractable sugar with existing manufacturing processes) was a major contention. His argument was that the effect of increased mineral content in the cane on extraction efficiency remained unanswered. Also, Thompson's studies produced inexplicable results—data so unusual that the entire basis of scientific agriculture was placed in doubt. For

38. Massachusetts Agricultural College, *Charles Anthony Goessman* (Cambridge, Mass., 1917), 37; C. A. Goessman, "Report on Recent Experiments with Sugar-Cane in Louisiana," *Proceedings of the American Chemical Society*, II (1879), 52–56; C. A. Goessman, "On Experiments with Fertilizers upon Sugar-Cane, at Calumet Plantation, Bayou Teche, La.," *Journal of the American Chemical Society*, I (1879), 416–20; Charles A. Goessman, *Notes on the Manufacture of Sugar in the Island of Cuba* (Syracuse, N.Y., 1865).

39. Bright, *New Orleans Price Current: . . . Crop Year 1876–1877*, XXVIII.

example, at the end of one year's trial, the results pointed to the fact that the best cane plat was that to which no fertilizer was applied to the soil at all! In a later test, the highest yield was gained by the application of gypsum alone, which, in the words of Goessman, was "a contradiction to the teaching of the best established experimental observations in agriculture."[40]

In the face of continuing confusion concerning fertilizers, reactionaries reveled. "I will tell him about the best 'fertilizers' I ever used on cane. I used it long before the war . . . ; since then it has become scarce and hard to obtain. It is thus: . . . drops of darky sweat, mixed with . . . 'elbow grease.' Applied up and down the rows every eight or ten days, will give the most . . . results; and if 'A Planter' will apply this fertilizer in heroic doses, he will never need any other preparation."[41] These sentiments may in fact have represented the majority view among Louisiana planters. It appears that it was only a small circle of progressive sugar planters who were attracted by the application of scientific methods to the solution of agricultural and manufacturing problems. The book that perhaps had the most impact upon this group of planters was L. Walkhoff's German treatise translated in French as *Traité Complet de Fabrication et Raffinage du Sucre de Betteraves* (2nd ed., 1875).[42] Indeed, several leading planters, including Duncan Kenner, John Dymond, John Wallis, and Joseph Godberry, claimed during the course of court testimony that they had read not only Walkhoff, but also such periodicals as the *Sugar Cane* and *Journal des Fabricants de Sucre*. French-authored treatises on sugar technology published during the 1870s continued to be written in the format of the 1840s, employing descriptions of processes and apparatuses and citing isolated examples of chemical analysis. In contrast, German workers rigorously presented process designs in conjunction with substantial analytical data. For example, in Walkhoff's work, large-scale processes were evaluated by the systematic use of analytical data. Thus, the filtration process and its efficiency were represented graphically by plotting saccharine

40. Goessman, "On Experiments with Fertilizers upon Sugar-Cane, at Calumet Plantation, Bayou Teche, La.," 419.

41. *Louisiana Sugar Bowl* (New Iberia, La.), April 29, 1875.

42. U.S. Treasury Department, *The United States vs. 712 Bags of Sugar Imported in the Mississippi*, n.p. [1878?], 23–24, 38–39, 44–45, 47, 53–54, 83, 90, 98.

concentration, alkalinity, and decoloration versus time. Walkhoff's treatise, as well as technical journals from France, Germany, and Britain, was an important source of knowledge for the sugar planters who would form the nucleus of the Louisiana Sugar Planters' Association in 1877. These planters realized that sugar manufacturing was becoming an industrial activity dependent upon scientific knowledge and engineering techniques. As a result, an important component of the newly formed association's energies was channeled into restructuring the Louisiana sugar industry to reflect recent European developments.

Both the use of the diffusion process for cane juice extraction and the scientific application of fertilizers underwent inconclusive trials during the 1870s. However, the question arising from the individual attempts to cultivate better quality cane and to extract a higher percentage of sucrose became research projects at the United States Department of Agriculture and the local sugar experiment station during the 1880s. Scientific and trade-related institutions that were emerging locally considered the solution of these problems, first recognized in the 1870s, to be future organizational objectives. While European ideas were crucial to the development of the Louisiana sugar industry, it was their assimilation within the framework of local institutions that ultimately proved crucial to the modernization of the industry.

4

Planters in Power
The Louisiana Sugar Planters' Association

The rise of a cluster of scientific, technical, educational, and business institutions proved to be at the heart of the late-nineteenth-century modernization of the Louisiana sugar industry, and these organizational developments were a direct response to challenges posed by a dynamic international market. Prior to the Civil War, Louisiana planters had little reason to band together and form organizations to promote their mutual interests, since prices were high, tariff rates more than adequate, and competition negligible. The antebellum period was generally a prosperous one for the sugar planter, and perhaps that is why such institutions as the University of Louisiana and the Agriculturists' and Mechanics' Association of Louisiana obtained little financial support from the community. It appears that, in the case of the Louisiana sugar industry, external pressures were a crucial ingredient in the formation of institutions.

One futile attempt to organize planters took place in 1856 when a sugar planters' convention was held in New Orleans to protest the handling of sugar and molasses on the levee.[1] Products were often allowed to sit beside the river for extended periods of time and when sold would bring extremely low prices. The convention appointed a committee to study this problem, and after deliberation it recom-

1. Sitterson, "Financing and Marketing the Sugar Crop of the Old South," 197–99.

mended the establishment of the Louisiana Sugar Mart to efficiently store and sell products. This institution did not become a reality until after the war; however, it remained until 1884 when another generation of planters, this time facing growing competition from Europe and elsewhere, created the Louisiana Sugar Exchange to handle sugar and molasses. As long as prices were high and demand was strong, planters overlooked inadequate facilities that existed within the state.

When the institutional framework of post-Reconstruction Louisiana crystallized after 1877, the resulting organizational matrix was in large part shaped by its central institution, the Louisiana Sugar Planters' Association (LSPA). Founded and led by many of the wealthiest and most politically powerful sugar planters in the state, the LSPA systematically developed alliances with federal government officials, practical engineers, and academic scientists to gain its organizational objectives. This group successfully lobbied for protective tariffs that ensured the continued rebuilding of the sugar industry devastated by the Civil War. Further, it persuaded the United States Department of Agriculture (USDA) to investigate sugar cultivation and manufacturing problems. As mentioned previously, in 1884 the association gained control of the marketing mechanisms in Louisiana by creating the Louisiana Sugar Exchange, and a year later it funded a private sugar experiment station in Kenner, Louisiana. By 1888 the LSPA was publishing its own weekly trade newspaper, the *Louisiana Planter and Sugar Manufacturer*, and in 1891 the association established the Audubon Sugar School for training experts for the sugar industry.[2] The establishment of the LSPA was crucial to the modernization of the sugar industry between 1880 and 1900, and the scientific and technological changes that occurred must be placed within this institutional context.

An analysis of LSPA origins, membership, and objectives is not only important because of the organization's significance to the modernization of the sugar industry, but also because the association reflected a dominant component of the social structure that existed in south Louisiana between 1880 and 1910. Since the publication of C. Vann Woodward's *The Origins of the New South* in

2. Sitterson, *Sugar Country*, 252.

1951, one major theme of scholarship has centered on the question of continuity in social relations from the antebellum to Reconstruction periods.[3] Woodward, like Henry Grady, maintained that a new middle class emerged after 1877 that participated in the economic rebirth of the South and contributed to its industrialization. Yet careful studies of business leaders in North Carolina contradict Woodward's view. For example, J. Carlyle Sitterson discovered that the industrial growth in that state, particularly in textiles, was directed by a group of businessmen who had their roots in the Old South, were closely tied to the plantation aristocracy, and often were educated at the University of North Carolina. These men used their position and its inherent advantages to further their own economic interests and thereby contributed to the state's prosperity.[4] Recent scholarship, especially the work of Dwight B. Billings, Jr., supports Sitterson's interpretations. Billings showed that three types of businessmen, all having ties to either the planter or merchant classes, had a powerful influence in shaping twentieth-century North Carolina.[5] In Billings' opinion, the New South was a revolution from above facilitated by the alliance of a weak middle class with a dominant planter aristocracy. Billings concluded that in terms of social relations the New South was characterized more by continuity with the past than by a radical restructuring of classes and power. In both Sitterson's and Billings' studies, carpetbaggers played a minor role in the new economic order.

However, developments in North Carolina cannot be extended to south Louisiana. An examination of the LSPA suggests that transplanted northerners, working in alliance with sugar planters of the old aristocracy, injected progressive views concerning science, technology, and business practices that were adopted by many prominent

3. Sheldon Hackney, "Origins of the New South in Retrospect," *Journal of Southern History*, XXXVIII (1972), 191–216.

4. J. Carlyle Sitterson, "Business Leaders in Post–Civil War North Carolina, 1865–1900," in Sitterson (ed.), *Studies in Southern History* (Chapel Hill, 1957), 111–21.

5. Dwight B. Billings, Jr., *Planters and the Making of a New South* (Chapel Hill, 1979). See also Jonathan M. Wiener, *Social Origins of the New South: Alabama, 1860–1885* (Baton Rouge, 1978), and Wiener, "Class Structure and Economic Development in the American South, 1865–1955," *American Historical Review*, LXXXIV (1979), 970–93.

planters, merchants, and manufacturers. Each group needed the other to survive economically. Those carpetbaggers who stayed through the 1880s and 1890s brought fresh ideas to Louisiana, but perhaps in an effort to gain acceptance, they took on the political and racial views of the old order. The interplay between northerner and native southerner, Republican and Democrat, Union and Confederate veteran, often took place at LSPA monthly meetings and frequently at the informal discussions held prior to these gatherings at the bar of the old St. Charles Hotel.

The Origins of the LSPA

Faced with high wage, interest, and transportation rates, south Louisiana planters began to organize during the early 1870s. Committees were established for the purpose of reducing both field-hand wages and railroad freight rates. In December, 1873, a call was made to formally organize "a state Grange . . . as soon as possible. By such action on the part of planters, it will reinspire confidence both among them and the common merchants. We find with the latter class, the opinion is universal that unless *something* is done to render planting less uncertain and more profitable, that it will not do to make further advances. Hence, the *only hope* left for planters is *to organize Granges at once.*"[6]

The Louisiana Grange movement contributed both leadership and ideology to future planters' organizations. Theodore Saloutos, a historian of the Grange movement in the South, has claimed that the large sugar planters of Louisiana were never attracted to the Grange movement.[7] However, contrary to Saloutos' assertions, several of the largest planters of St. Mary Parish not only provided the leadership for the Grange's central committee, but also guided the formation of a Grange committee for the reduction of freight rates. And at least one of the Grange's leaders, Charles Walker, a transplanted midwesterner and Bayou Teche planter, later became a vice-president of the LSPA.

In particular, the Grange's interest in the quality of education in

6. *Louisiana Sugar Bowl* (New Iberia, La.), December 18, 1873.
7. Theodore Saloutos, "The Grange in the South, 1870–77," *Journal of Southern History*, XIX (1953), 473–87.

the state and the responsiveness of the agricultural and mechanical college to local needs became concerns of future planter associations.[8] Although the Grange's attempt to control the mismanaged agricultural and mechanical college failed, planters began to recognize that local economic growth depended upon the existence of scientific and technical education. An 1873 essay in the *Louisiana Sugar Bowl* stated: "Manual labor is the wealth, the power which explains the character of a people. . . . We must have mechanical education, industrial schools, we must make mechanics of our young men." In addition to the need for mechanical schools, one Lafourche Parish planter, acquainted with "the best scientific schools of the country," asserted in 1874 that Louisiana State University cadets must be taught "a scientific education, thorough and complete."[9]

In general, with the exception of the formation of a few local planters' groups and their minor participation in the Grange movement, Louisiana planters remained unorganized on a statewide basis throughout the Reconstruction era. One local association was the Teche Planters' Club, which held its first meeting in Franklin, Louisiana, in October of 1876. The editor of the *Louisiana Sugar Bowl* viewed the club as a kind of county agricultural association, which in facilitating the interchange of information would consequently develop scientific agriculture as a body of knowledge. The ostensible purpose of the organization was "of elevating agriculture to the standard of a *science* and *Profession*." However, the organization never developed a scientific and technical focus. Club discussions covered a diverse number of topics, including chemical fertilizers, the introduction of northern labor, and corn growing in the South.[10]

It appears that while planters had lofty intentions about introducing science on the sugar plantation, their immediate concern with the wages of field workers, transport costs, and interest rates predominated the monthly meeting's agenda. Indeed, the most pressing problem facing Louisiana planters at the time was not so much the unfair business practices of the "monster" railroad, but how to secure an adequate and relatively efficient unskilled work force to

8. *Ibid.*, 486.

9. *Louisiana Sugar Bowl* (New Iberia, La.), May 22, 1873, and April 23, 1874.

10. *Louisiana Sugar Bowl* (New Iberia, La.), July 27, 1876, October 26, 1876, and November 19, 1876.

cultivate, cut, and haul sugarcane. It was in dealing with the unsettled labor situation that the planters first recognized the power of local organizations. Throughout the 1870s planters, at times acting in concert, ironically attempted to improve the morale of black field hands while somehow keeping wages as low as possible. In this way daily salaries were established in the sugar bowl. Not only was there a movement to reduce and stabilize wages, but also an effort was mounted to replace black workers with Chinese, Swedes, and Italians, an experiment that met with little success. As one overseer remarked, "I think it bad enough to have drunken niggers but worse to have drunken Italians."[11]

Nowhere was this concern with labor more evident than in the activities of a planters' organization that was formed in St. Mary Parish in 1877. The group resolved to limit the wages of sugarhouse workers, as well as to change the terms of payment for field laborers. To insure that a constant supply of labor remained available throughout the entire harvest season, the planters agreed that monthly wages would no longer be paid, but rather one-third of the total wages owed to unskilled labor would be withheld until the sugarcane season ended.[12]

Apparently, the Teche Planters' Club ceased to hold its monthly meetings after November, 1877. During that same month an organization different in both membership and objectives formed in New Orleans. While the Teche club had only a local focus, the newly formed Louisiana Sugar Planters' Association (LSPA) centered its initial efforts on securing federal legislation favorable to the Louisiana sugar industry. This initial political success provided the confidence necessary to seek additional power by other means. The leadership recognized that the application of scientific methods and new technological innovations to a traditional industry would enhance their economic status not only within the local community, but also in the world market. However, the leaders of this statewide planters' association did not totally ignore the everyday problems confronting

11. Richard J. Amundson, "Oakley Plantation: A Post–Civil War Venture in Louisiana Sugar," *Louisiana History,* IX (1968), 31. See also J. Vincenza Scarpaci, "Labor from Louisiana's Sugar Cane Fields: An Experiment in Immigrant Recruitment," *Italian Americana,* VII (1981), 19–41.

12. *Louisiana Sugar Bowl* (New Iberia, La.), November 11, 1877.

Louisiana sugar planters. During the 1880s local affiliates of the LSPA tackled the labor, capital, and transportation problems first addressed in the 1870s by the Grange, the Teche Planters' Club, and the planter group in St. Mary Parish.

Organizers of the Louisiana Sugar Planters' Association first attempted to create a formal organization in 1870. In the spring of that year, a number of planters, led by Duncan Kenner from Ascension Parish, met at the St. Charles Hotel in New Orleans. Among those present were Kenner, Dr. J. B. Wilkinson of Plaquemines Parish, Andrew McCollam of Terrebonne Parish, and Judge A. B. Merrill, manager of a plantation in Jefferson Parish. While it was planned that future meetings would be held on a quarterly basis, the first formal meeting, scheduled for October, 1870, never took place, apparently because of the outbreak of yellow fever in New Orleans.[13] No meetings were held during the next seven years, until Kenner, perhaps urged on by John Dymond, again attempted to form a planters' group. On November 20, 1877, seventeen men closely associated with the Louisiana sugar industry met at Kenner's office to discuss the impending tariff legislation which called for a downward rate revision on sugar. According to the minutes of that first LSPA gathering, the Louisiana sugar interest organized because of proposed changes in the tariff that would lower the duty on sugar. Planters had "to take steps for their better protection."[14]

Although several printed sources tend to indicate that Kenner played a key role in the founding of the LSPA, John Dymond, a future president of the organization and editor of the *Louisiana Planter and Sugar Manufacturer*, recalled some thirty-two years after the association's establishment that his own efforts were crucial.[15] According to Dymond, since Duncan Kenner had close ties to the antebellum aristocracy in Louisiana, Kenner was persuaded to assume

13. Louisiana Sugar Planters' Association, *Regular Monthly Meeting, July 1887* (New Orleans, 1887), 3.

14. Louisiana Sugar Planters' Association, "Minutes of Monthly Meetings, 1877–1891" (MS in Special Collections, Hill Memorial Library, Louisiana State University, Baton Rouge, La.), 1, 3.

15. "Louisiana Sugar Planters' Association: Presentations to the Association of the Portraits of Etienne de Bore and the Past Presidents of the Association," *LP,* XLII (1909), 59–62.

the presidency of the LSPA, thus providing the association with leadership that was socially and politically acceptable to the majority of the state's planters. Dymond, a midwestern carpetbagger with New York trade connections, secured the signatures of prominent planters who agreed to join the LSPA and worked tirelessly behind the scenes to establish an institution that would initially be led by Kenner and controlled by a circle of planters and merchants having their roots in the Old South.[16]

In 1909 Dymond reminisced:

> The matter was under consideration for several years and up to 1877. We began to feel the pressure rather severely of competing with one another and I suggested to some of our friends that we should endeavor again to effect organization, and in New York I was discussing the matter with leading sugar brokers there, and one day Francis Skiddy [a broker and uncle of Dymond's business partner] came to me and said, "if you want to make your organization in Louisiana, you had better make it right away." I told him I certainly would. We arranged a meeting that afternoon with two or three prominent sugar refiners including the leaders of the Sugar Trust, and we began negotiations concerning a tariff that would satisfy us in Louisiana as well as them in New York.[17]

As a result of the meeting, a statement of the purposes and objectives of the association was published on November 29, 1877, in the *Louisiana Sugar Bowl*. The organization was a response to the "disastrous situation of the sugar interests of Louisiana." It was hoped that an alliance, not unlike that of unions, guilds, or associations for advancing professional or trade interests, would harmonize and concentrate the sugar interest. The association would be similar to other groups, "whose purpose it is to foster the interests of the class they represent, . . . and they have been found powerful and efficient agents in securing legislation, or other aid in behalf of their constituencies." The leaders asked, "Shall we, as a class, remain idle and indifferent, totally unorganized, trusting only to individual effort to protect our interests, or shall we, by united and concentrated asso-

16. "Biography of John Dymond" [MS in Dymond Family Papers, Folder 338, HNOC], 2.
17. "Louisiana Sugar Planters' Association: Presentations to the Association of the Portraits of Etienne de Bore and the Past Presidents of the Association," 61.

ciation, give greater force and efficiency to our action?"[18] The association's objectives were to "develop the culture of sugar cane, the manufacture of sugar therefrom in all its branches, to furnish statistics and facts as well, justify favorable legislation on the part of the United States Congress in behalf of this great industry, and to harmonize and concentrate . . . , the efforts of all those engaged in the cultivation, manufacture and handling of the sugar products of this State, and of those who are engaged in the manufacture of machinery."[19]

The Louisiana Sugar Planters' Association was led by its president, "whose duty it was to preside at all meetings . . . , and exercise a general supervision over its affairs."[20] The president was also an ex officio member of all committees, including the standing executive and finance committees. The president, the three vice-presidents, a secretary, a treasurer, and the five members constituting the executive committee were to be elected annually at the March meeting. While the finance committee was responsible for the finances of the association, important affairs were always managed by the executive committee. As the situation demanded, other committees were formed to evaluate new technology, to investigate the formation of a sugar exchange, to contact university authorities, and to gather statistical information. Although the president selected the chairman and members of each temporary committee, both the LSPA membership and the president could initiate a request for the formation of a special investigative group. After the committee had accomplished the assigned task, the committee chairman normally presented a report at the monthly meeting. It was then determined if the committee's recommendations were to be accepted or whether further research or other actions were advisable.

In many respects the objectives of the LSPA were no different from those of organizations for beet-sugar manufacturers in Europe. Sugar producers in Vienna and Prague had formed an association to gather relevant statistical data, to introduce harmonious commercial practices, to establish common marketing warehouses, and to negotiate

18. *Louisiana Sugar Bowl* (New Iberia, La.), December 13, 1877.
19. "Constitution" (MS in Dymond Family Papers, Folder 231, HNOC), 1.
20. *Ibid.*, 1–5.

with government authorities for favorable tax laws.[21] While Louisiana planters were apparently aware of European developments in the beet-sugar industry in 1877, there is no evidence to suggest that the LSPA was consciously modeled after a similar European organization. Rather, the formation of the LSPA was a response to the impending threat of tariff reductions. Once they were assured that favorable sugar duties would continue, the LSPA leaders also recognized that the Louisiana sugar industry's long-term survival depended upon the ability of its planters to adopt and apply scientific methods and advanced engineering techniques.

The Members: Planters and Merchants

Although not all of the large producers joined the Louisiana Sugar Planters' Association, its membership included a representative group of the wealthiest landowners and largest sugar manufacturers in the state. While the average sugar producer in Louisiana made 153 hogsheads during 1879–1880, the production of key LSPA leaders was much greater. Duncan Kenner made 805 hogsheads; Edward J. Gay, 1,881; the McCall family, 875; Richard Milliken, 1,018; Adam Thomson, 825; Bradish Johnson, 1,924; and John Dymond, 1,323.[22] Yet their common economic interest was not the only cohesive force uniting the members. Often other concerns and bonds were just as important. A complex matrix of interpersonal relationships existed, including social club affiliations, kinship, local neighborhood connections, and allegiance to national political parties.

During the 1880s an alliance based solely upon economic interests developed between opposing political factions within the sugar interest. Republicans were led by H. C. Warmoth, and Democrats looked to Duncan Kenner and William Porcher Miles for leadership. This union was purely a marriage of convenience, and the LSPA became the common meeting ground for these two disparate groups. The relationship proved fruitful for both planter factions, as well as the Louisiana sugar interest in general. From the ranks of the Demo-

21. *Statuten des Assekuranzvereins Oesterreichicher Zuckerfabrikanten in Prag* (Prag, 1862), 7.
22. A. Bouchereau, *Statement of the Sugar and Rice Crops Made in Louisiana in 1879–80* (New Orleans, 1880), xvi.

cratic planters came a number of experienced politicians familiar with legislative processes in Washington. Although also active in political negotiations, Republican planters were particularly influential in promoting the adoption of scientific methods and advanced technology in the Louisiana sugar industry. Delegations lobbying for the LSPA in Washington between 1879 and 1890 were typically composed of members of both political parties. For example, in 1880 Judge Taylor Beattie, a Republican planter from Terrebonne Parish, traveled to Washington with Democrats Kenner and Dymond to campaign against tariff reductions. In 1886 Warmoth, along with Democrats Dymond and Henry McCall, promoted LSPA interests by visiting John Sherman, the president of the United States Senate.[23] Yet apart from attending monthly LSPA meetings and joining together for the common cause of ensuring the continuance of favorable tariff legislation, members of these two groups rarely interacted socially.

Many early Democratic members of the LSPA were also members of the exclusive Boston Club in New Orleans. In its handbook, the social club described itself as "a world within itself." While the club was originally organized during the 1840s to promote the card game Boston, by the early 1890s it had taken on an air of respectability by declaring its primary purpose to be "the cultivation of literature and science."[24] Of the thirteen men who served as the first officers and members of the LSPA standing committees, seven, including Kenner, Thomas Miller, Emory Clapp, Adam Thomson, Samuel Kennedy, and Richard Milliken, were members of the Boston Club prior to 1877, and three more joined after the LSPA was founded.[25] The membership of the Boston Club also included a number of other LSPA members who, though not in the original leadership circle, eventually played a prominent role in fulfilling the group's objectives. These members included Joseph L. Brent, Louis Bush, Randall Lee Gibson, Bradish Johnson, Harry and Richard McCall, Emile Rost, and W. B. Schmidt. All of these Democratic planters and merchants

23. "Delegation to John Sherman, President of the Senate," March 8, 1886, in Warmoth Papers, Reel 4, SHC.

24. *Officers, Members, Charter and Rules of the Boston Club* (n.p., 1966), 4; *Charter and Rules of the Boston Club of New Orleans* (New Orleans, 1892), 5.

25. Stuart O. Landry, *History of the Boston Club* (New Orleans, 1938), 220–59.

in the Boston Club either had served as officers in the Confederate army or had family ties to the antebellum planter class.[26] Thus it was natural for Kenner, a former member of the Confederate Congress, to propose Randall Lee Gibson, a former Confederate military officer, for membership in the club.[27] With its membership firmly rooted in the plantation culture of the Old South, the Boston Club did not welcome carpetbag sugar planters. John Dymond—a midwesterner born of English parents, and a staunch Democrat from the time of his arrival in Louisiana in 1867 until his death in 1921—was the only carpetbag sugar planter to be admitted to the Boston Club, and then not until 1911.

While the social club was one focus of common interest, kinship ties were also important within the LSPA.[28] Duncan Kenner, the son of a prosperous merchant, was born in New Orleans in 1813. After receiving his early education from private tutors, he enrolled at Miami University of Ohio, from which he graduated in 1831.[29] After a period of European travel, Kenner read law with John Slidell. Rather than practice law, however, Kenner became a sugar planter in Ascension Parish. In 1839 he married into an important Louisiana Creole family, the Bringers.

Family ties like those of Kenner's often played an important role in the adoption of new technology in the sugar bowl. For instance, Kenner often gave advice to relatives like William J. Minor concerning the dimensions of his steam engines and the best supplier of mechanical equipment.[30] And the gradual acceptance of John S. McDonald's hydraulic pressure regulator was in large part due to trials conducted on the Bringer, Turnead, and Kenner plantations, all owned by close relatives.

Kenner began his political career in 1836 when he was elected

26. C. Lynch Hamilton to William Porcher Miles, February 22, 1894, in William Porcher Miles Papers, Box 5, Folder 70, SHC.

27. "Election Book 1885–1888" (MS in Boston Club Collection, Manuscripts Department, Special Collections Division, Tulane University), 6.

28. Stanley C. Arthur and George de Kernion, *Old Families of Louisiana* (New Orleans, 1931), 157–60, 362–64, 428–30; *Biographical and Historical Memoirs of Louisiana* (Chicago, 1892), I, 27–72, 352, 519, 535, 545, and II, 259–60, 323–24, 342.

29. *Dictionary of American Biography*, V, 337–38.

30. Duncan F. Kenner to William J. Minor, January 22, 1846, in Duncan Kenner Papers, Box 1, Folder 7, Special Collections, Hill Memorial Library, LSU.

to the Louisiana House of Representatives. He later served several terms in both the House and Senate. In 1861 he was a member of the Provisional Congress of the Confederacy, and he later became a Louisiana state representative in the Confederate Congress, where he eventually became chairman of its Ways and Means Committee.

As the Civil War progressed, Kenner became convinced that the success of the Confederacy depended upon European recognition and that slavery hindered that goal. In 1864 Kenner persuaded Judah P. Benjamin, the secretary of state of the Confederacy, to send him on a special mission to Britain and France. The scheme, which was approved by Jefferson Davis, called for the abolition of slavery in return for national recognition. In disguise and using the name of A. B. Kingslake, Kenner penetrated Union lines and sailed from New York on February 11, 1863.[31] Hearing reports of Sherman's march, European diplomats clearly saw the eventual demise of the Confederacy, and Kenner's mission ended in failure. Upon his return in late 1865, Kenner found his plantation in ruins, but undaunted by the South's defeat, he rebuilt it larger than it was before the war. To a great extent Kenner's post–Civil War financial success was due to his civic boosterism that led him to promote an improved canal system for New Orleans, to manage the Louisiana Levee Company, to work for the chartering of the Metairie Cemetery, and to sell New Orleans city bonds.[32] Kenner was also actively engaged in New Orleans real estate activities and perhaps invested in the Tehuantepec Inter-Ocean Railroad in Mexico, a railway that promised to increase the commercial clearings in the port of New Orleans.[33] At the time of his death, Kenner held large amounts of stock in the New Orleans Gaslight Company, Jefferson City Gaslight Company, the Louisiana Sugar Exchange, the New Orleans Pacific Railroad, and the American Sugar Refining Company.[34] Kenner proved to be not only a skillful businessman but also an astute politician. He was a state senator between

31. "List of Passengers of the Bremen Steamship 'America,'" February 11, 1865, in Kenner Papers, Box 1, Folder 2, LSU.

32. Duncan Kenner to Withers, May 16, 1872, in Kenner Papers, Box 1, Folder 4, LSU.

33. N. B. Trist to D. F. Kenner, June 17, 1877, in Kenner Papers, Box 1, Folder 5, LSU; *Tehauntepec Inter-Ocean Railroad, Mexico* (n.p., 1882).

34. "Succession of Duncan F. Kenner" (MS in Kenner Papers, Box 1, Folder 9, LSU).

1866 and 1867 and again in 1877. During the Hayes-Tilden controversy in 1876–1877, he was in Washington, representing Louisiana
interests. In 1878 he ran for the United States Senate and lost, but he
was appointed by President Chester A. Arthur to the United States
Tariff Commission in 1882.

Kenner's relatives and close neighbors in the Louisiana Sugar Planters' Association provided him with valuable support during his term
as president of the organization. His brother-in-law and business
partner, Joseph Brent, was a strong proponent for the introduction of
scientific methods into the Louisiana cane-sugar industry. Brent was
born in Baltimore and educated at Georgetown College.[35] After a
brief law career in California during the 1850s, Brent served as a
Confederate cavalry officer. Shortly after the war he returned to
Baltimore to resume his legal career. In 1870, after his marriage to
Rosella Kenner, he moved to Louisiana, where he became a successful sugar planter and politician. He served as the first president of
the Louisiana Agricultural Society and was elected to two terms
in the Louisiana State Legislature.

Kenner's neighbors in Ascension Parish and nearby Lafourche
Parish, particularly the McCall family and William Porcher Miles,
supported the early LSPA objectives with participation and financial
support.[36] The McCall family owned several large plantations across
the Mississippi River from Kenner's holdings; the title to a portion of
this land dated from the 1780s. Henry McCall was born in 1847 in
Louisiana. He received his preliminary schooling in Maryland and
completed his education in France and Britain.[37] In 1867 he returned
to New Orleans where he went to work as a cotton broker. Two years
later he took over the management of his family's Evan Hall plantation, which then consisted of some 2,500 acres of sugarcane under
cultivation. Henry McCall was a most active participant in the

35. *Appleton's Cyclopedia of American Biography* (1900), VII, 33; Brent wrote
several memoirs, including *The Lugo Case* and *Capture of the Indianola* (New Orleans, 1926), and *Memoirs of the War between the States* (New Orleans, 1940).

36. Neighbor relationships were established by examining A. Bouchereau's *Statement of the Sugar and Rice Crops Made in Louisiana* and William J. Hardee, *Hardee's
New Geographical, Historical and Statistical Official Map of Louisiana* (Chicago,
1895).

37. *Biographical and Historical Memoirs of Louisiana*, II, 214.

monthly LSPA meetings throughout the 1880s, often reading papers on the application of new technology to sugar manufacturing. Like many of the leading figures in the association, McCall was also active in community affairs and politics, serving during the 1880s as president of the Ascension Parish Police Jury and on the Board of Supervisors of the Agricultural and Mechanical College of Louisiana. Miles, another neighbor of Kenner's, was a former mayor of Charleston, South Carolina, former Confederate legislator, and past-president of the University of South Carolina. He came to Louisiana in 1882 to manage the Ascension Parish plantation of his uncle, Oliver Bierne. Miles later assumed ownership of the plantation, and by 1890 he had become the largest single sugar producer in Louisiana.

In 1884 Miles, Brent, Henry McCall, and James P. Kock formed the nucleus of the Ascension Branch of the Sugar Planters' Association (ABSPA).[38] James Kock, born in 1853 in Louisiana, was educated in Switzerland, France, and Germany before returning to the United States in 1869. After serving for several years as an accountant for a German import company in New Orleans, Kock assisted his brother in the management of Belle Alliance plantation near Donaldsonville, taking complete control of the operation in 1884. Kock faithfully attended meetings of the ABSPA, which were first led by Brent and later by Miles, but which did not have the scientific or technical focus of the LSPA. Although papers were occasionally read by ABSPA members, the main function of the organization was to enable neighboring planters to dine together and exchange information concerning labor problems and politics. In particular, the ABSPA provided planters with the opportunity to establish uniform wage rates through written agreements.[39] Also, throughout the 1880s the Ascension Parish sugar planters' organization served as one focal point for conducting Democratic party campaigns in Louisiana. In 1884 Miles, with the support of Brent, urged the group to adhere "to Dem. Party & [emphasized] the vain expectations of those who would ally them-

38. William Porcher Miles Diary, May 7, 1884, in William Porcher Miles Papers, Vol. 18, SHC.

39. Miles Diary, July 27, 1884, and October 2, 1884, in Miles Papers, Vol. 18, SHC; Joseph L. Brent to William P. Miles, December 13, 1886, in Miles Papers, Box 4, Folder 56, SHC.

selves (in any way) with Rep. party, because [it was] the party of protection." Thus, in 1884 Miles, McCall, and other Ascension Parish planters actively campaigned for the election of Edward Gay to the United States Congress.[40]

Edward James Gay was born in Virginia in 1816 and later moved with his family to Illinois and then to Missouri, where he attended Augustana College in 1833–1834. He was engaged in commercial activities in St. Louis until 1850. After the Civil War he moved to Louisiana, where he purchased plantations in several of the sugar parishes. Gay, a Democrat, was named the first president of the Louisiana Sugar Exchange in 1883 and was subsequently elected United States congressman representing Louisiana between 1885 and 1889. His son-in-law, Andrew Price, graduated as a lawyer, then managed Gay's plantations during the 1880s. Eventually Price succeeded Gay in the House of Representatives. After his election in 1889, Price remained in close contact with the Ascension Parish planters, and he corresponded with Miles concerning manufacturing expenses and tariff legislation.[41]

Along with Gay and Price, Senator Randall Lee Gibson was another politician whose presence in Congress was crucial during early LSPA efforts to maintain high duties on sugar.[42] Gibson was born in 1832, educated by private tutors at his father's Live Oak plantation in Terrebonne Parish, and graduated from Yale College in 1853. He obtained a law degree from the University of Louisiana in 1855, and after a European tour returned to agricultural pursuits in Louisiana. During the Civil War he rose to the rank of brigadier general in the Confederate army and later returned to Louisiana to practice law and rebuild his plantation. After an unsuccessful campaign in 1872, he

40. Miles Diary, June 4, 1884, in Miles Papers, Vol. 18, SHC; E. N. Pugh to W. P. Miles, October 30, 1884, in Miles Papers, Box 4, Folder 55, SHC.

41. *Biographical Dictionary of the American Congress, 1774–1971* (Washington, D.C., 1971), 991; "Andrew Price," *LP*, XLII (1909), 114–15; Miles Diary, March 8, 1892, in Miles Papers, Vol. 26, SHC; Andrew Price to William P. Miles, August 31, 1893, in Miles Papers, Box 5, Folder 68, SHC.

42. *Biographical Dictionary of the American Congress 1774–1971*, 999. *Senate Documents*, "Memorial Addresses on the Life and Character of Randall Lee Gibson," 53rd Cong., 2nd Sess, No. 178.

was elected as a Democratic congressman to three successive congresses (1875–1883), and then served as a United States senator from 1883 until his death in 1892.

During the 1880s Gibson's neighbors in Terrebonne Parish organized a local planters' organization much like the ABSPA in Ascension Parish. On December 23, 1886, a meeting was held in Houma, Louisiana, "for the purpose of organizing an association to be composed of representative men of every industry, trade or profession, to conserve the interest of the parish of Terrebonne."[43] The secretary of the newly formed group was requested to obtain a copy of the bylaws and constitution of the ABSPA, and it was stated that the group would address problems concerning railroad freight costs, interest rates, and brokerage and handling charges for sugar transported to New Orleans. The concerns first raised by the Grange movement in the 1870s had become central topics of interest for local sugar planters' associations in Louisiana in the 1880s.

In addition to Ascension and Terrebonne parishes, Plaquemines Parish was a third major sugar-producing area in Louisiana, and several planters from this area were important LSPA members. Henry P. Kernochan, first vice-president of the LSPA, was closely tied to numerous other members of the organization by kinship, prior Confederate military service, and close-neighbor friendships.[44] Kernochan was a New Yorker who came to Louisiana before the Civil War. He fought for the Confederacy under General Randall Lee Gibson, and after the war Kernochan purchased the old Packwood plantation in Plaquemines Parish, which he managed from 1866 to 1895. He was a member of the Louisiana State Convention of 1879 and in 1885 was appointed naval officer for the Port of New Orleans. Kernochan's nephew, George Garr, was also an active LSPA member. Garr's father had owned a Louisiana plantation before the war, and the New York family had held the plantation as a part of a larger commercial enterprise.[45]

After the Civil War, several of Kernochan's neighbors developed strong interests in the application of science and technology to sugar

43. "Sugar Planters," December 31, 1866 (Unidentified newspaper clipping in William A. Schaffer Papers, Scrapbook, Vol. 5, SHC).

44. "Henry P. Kernochan," *LP*, XXX (1903), 312.

45. "Joseph K. Garr," *LP*, XLI (1908), 259.

manufacturing problems and information was often exchanged. In 1879 Kernochan provided information on his and his nephew's manufacturing expenses to planter Henry Clay Warmoth. And that same year Warmoth received advice on steam engine repairs from neighbor John Dymond.[46]

Within a few miles of Kernochan's properties were the Fairview and Belair plantations, owned by John Dymond. Without doubt Dymond was the most important individual contributing to the late-nineteenth-century modernization of the Louisiana sugar industry. He was a masterful negotiator in the political arena and an entrepreneur who recognized that scientific knowledge was power and that science was best applied to industrial problems through institutional means. Dymond was born in Canada in 1836, but he grew up in Zanesville, Ohio, where his father was a Methodist preacher and businessman.[47] In 1857 he graduated from Bartlett's College, a commercial school located in Cincinnati, and first worked in the mercantile business and later in cotton manufacturing. By the early 1860s Dymond had moved to New York, where he entered into the sugar and coffee import business with business partner James Lally. With the conclusion of the Civil War, Dymond and his associate purchased Belair and Fairview plantations in Louisiana, and the former's career as a sugar planter began. By the mid-1870s Dymond had bought out Lally and had begun to increase his Louisiana properties by lending money to cash-poor neighbors and subsequently obtaining their lands through forfeiture. Dymond was a carpetbagger, but he was a Democrat; and unlike many other northern businessmen working in the South, he was gradually accepted by native Louisianians. His political philosophy was similar to that of many southerners in that he claimed to adhere "to the state's rights doctrine and home rule for the white race."[48] In 1888 he was a delegate to the Democratic national convention and later served several terms as state senator.

A number of other Plaquemines Parish planters also played impor-

46. "Account Book, Miscellaneous Records 1877–1879" (MS in Henry Clay Warmoth Papers, SHC).

47. *National Cyclopedia of American Biography*, XXV, 350. "John Dymond" and "Biography of John Dymond" (MSS in Dymond Family Papers, Folder 338, HNOC).

48. "Biography of John Dymond," 8.

tant roles in the LSPA, including Bradish Johnson. Johnson was born in Louisiana in 1811.[49] He enrolled in and graduated from Columbia College in New York, where his father was engaged in the distillery trade. Later Johnson, too, entered that field. However, shortly before the Civil War, he inherited a sugar plantation. Johnson was among the first planters to use double mills and bagasse furnaces. He also developed close relationships with the White, Lawrence, and Wilkinson familes.

The Wilkinsons, descendants of General James Wilkinson, were a prominent antebellum Louisiana family. Dr. Joseph Biddle Wilkinson (1817–1902) received his education at the University of Virginia and the University of Pennsylvania. Apparently recognizing the changing nature of the world sugar trade, he was an early advocate of the planters' organization. It appears that Wilkinson was best known by contemporaries for his political power—a man who did not seek the limelight but whose "influence was felt in every direction." Wilkinson's political ambitions were realized through the career of his son, Theodore Stark Wilkinson (1847–1921). T. S. Wilkinson graduated from Washington College in Lexington, Virginia, in 1870, then returned to Louisiana to assist his father at the Point Celeste plantation.[50] Between 1887 and 1891 Wilkinson served in the United States Congress. After refusing nomination for a third term, he was appointed collector of the Port of New Orleans in 1893, replacing the Republican appointee, Henry Clay Warmoth.

While the Democratic planters in Plaquemines Parish contributed greatly to the LSPA efforts during the nineteenth century, Republican planters like Henry Clay Warmoth (1842–1931) also had important roles in the organization. Born in Illinois, Warmoth began as an apprentice typesetter and later became a lawyer in Missouri.[51] A for-

49. "Bradish Johnson," *LP,* IX (1892), 350; "Portraits of Prominent Planters," *LP,* XIV (1910), 185.

50. "Dr. Joseph Biddle Wilkinson," *LP,* XXIX (1902), 50–51; *Biographical Dictionary of the American Congress 1774–1971,* 1924.

51. Francis Byers Harris, "Henry Clay Warmoth, Reconstruction Governor of Louisiana," *Louisiana Historical Quarterly,* XXX (1947), 523–63; Francis Wayne Binning, "Henry Clay Warmoth and Louisiana Reconstruction" (Ph.D. dissertation, University of North Carolina at Chapel Hill, 1969); J. Carlyle Sitterson, "Magnolia Plantation, 1852–1862," *Mississippi Valley Historical Review,* XXV (1938–39), 197–210.

mer Union soldier, he practiced law in New Orleans after the Civil War and was elected governor of Louisiana in 1868. His term was characterized by discontent, turbulence, and corruption. In addition to politics, Warmoth was also actively engaged in business pursuits in Reconstruction Louisiana. During a time when capital was scarce, Warmoth had money to invest and shrewdly lent funds to friend and political ally Effingham Lawrence, who had inherited Magnolia plantation in 1855 from his father-in-law. Warmoth became a part owner of this large sugar plantation forty-six miles below New Orleans, and upon Lawrence's death in 1879, he came into sole possession of the property. At Magnolia plantation he began experimenting in new methods of sugar production and soon became a leading advocate of the new technology. With a letter of introduction from President Chester A. Arthur, Warmoth traveled to Europe in 1884 to investigate the European beet-sugar industry. After his return, the USDA established a sugar experiment station at his Magnolia plantation.

Despite Warmoth's controversial term as governor, he remained politically active in Louisiana state politics. In 1876–1877 he was a member of the Louisiana legislature, and in 1879 he served on the state constitutional convention. In 1888 he headed the state's Republican ticket but lost the election. He was then appointed collector of customs in 1890, a position he held until 1893. In a very real sense, Warmoth's ideas concerning politics and business cannot be separated. As a much-maligned political figure, Warmoth preached a gospel of business efficiency and economic growth, and he put these ideas into practice on his plantation. He advocated a protective tariff for the Louisiana sugar interest and during the campaign of 1888 proclaimed, "The great republican party—the solid businessmen of the North—will . . . help make Louisiana a great State, and New Orleans the great commercial entrepot it should be."[52] One wonders what would have become of the state had he been victorious at the polls.

Warmoth's interests in both Republican party politics and advanced sugar technology were shared by a group of transplanted midwesterners who settled along Bayou Teche after the Civil War. John B. Lyon,

52. "Fair-Minded Men Stop and Think," March 6, 1888, in Levert Family Papers, Louisiana State Museum.

a Chicago businessman, persuaded a number of midwestern merchants—Charles H. Walker, Daniel Thompson, Lewis S. Clarke, Dr.
Henry T. Landers, T. J. Bronson, John Foos, and James W. Barnett—to
invest in Bayou Teche plantations. Lewis S. Clarke (1837–1906) was
among the most prominent planters along the Teche and one of the
first to adopt the diffusion process in Louisiana at his Lagonda plantation. Active in Republican politics, he managed John N. Pharr's
1894 gubernatorial campaign, and in 1900 he was made Republican
national committeeman for Louisiana. Another midwesterner on
the Teche was James W. Barnett, originally from Springfield, Ohio.
Barnett operated the Shadyside plantation in St. Mary Parish with
his partner, John Foos.[53]

As a consequence of his innovative practices in business procedures and in the employment of scientific methods in sugar manufacturing, Daniel Thompson was among the most important of the
midwesterners who relocated to Louisiana after the Civil War.[54]
Thompson, owner of Calumet plantation, began his career as a manager of street railways and a trader on the Chicago grain market. During the 1860s he initiated a systematic study of sugarcane fertilizers
and later examined problems related to steam drainage, levee construction, and new apparatuses and processes. His ability to organize
and manage was reflected in the subsequent technical reports published at the Calumet plantation.

These midwesterners recognized that the post–Civil War plantations were inexpensive and that large returns on their capital investments were possible. Later John Dymond asserted that "this group
of men did probably more to advance cane culture and sugar manufacture as an industrial art in Louisiana than did any other group of
men associated with the industry."[55] While James Barnett never actively participated in Republican politics in Louisiana—he even committed himself to Democrat Andrew Price in 1889—Landers, Foos,
Thompson, and Clarke supported Republican candidates during the

53. "Lewis S. Clarke," *LP*, XXXVII (1906), 2–3; "James W. Barnett," *LP*, XXXIII
(1904), 117.

54. "Daniel Thompson," *LP*, XXIV (1900), 146. Hubert Edson, *Sugar—From Scarcity to Surplus* (New York, 1958), 46.

55. "Lewis S. Clarke," 2.

1880s and 1890s.[56] Yet economic interests often overshadowed partisan party politics. Republicans Thompson, Clarke, Warmoth, along with Democrats Henry McCall and James P. Kock, formed the core group of a select sugar club that apparently exchanged financial statements and technical information during the early 1890s.[57]

A number of LSPA members were merchants as well as planters. Richard Allan Milliken was born in Ireland in 1817 and educated during the 1830s at Bardstown College in Louisville, Kentucky.[58] At the beginning of the Civil War, Milliken factored about one-third of the Louisiana sugar crop. Although he was the commissioner of the Burra Copper Mines near Knoxville, Tennessee, during the war, he returned to Louisiana after the conflict to resume his business. As a consequence of his success as a sugar broker, Milliken eventually came to own several large sugar plantations in the state. James A. Murphy, who began as an errand boy for Milliken, took over the business in 1896.[59] Murphy became president of the Louisiana Sugar Exchange and went into partnership with Milliken's nephew, H. B. Farwell.

Another sugar factor prominent in LSPA activities was Louis Bush.[60] Bush was born in 1820 in Iberville Parish and practiced law in Lafourche Parish before the Civil War. During the conflict he was an officer in the 18th Louisiana Regiment and saw action at Shiloh. In 1872 he opened a commission business in New Orleans and later formed a partnership with John Baptist Levert, a former Confederate soldier and a St. Martin Parish planter.[61] Levert, who became president of the Louisiana Sugar Exchange during the 1880s, was born in 1841 and educated at Mount St. Mary's College in Emmittsburg, Maryland. After the war Levert returned to Louisiana and, along

56. John Foos to Henry Clay Warmoth, August 21, 1889, J. W. Barnett to Warmoth, May 22, 1892, J. B. Lyon to Warmoth, August 21, 1889, all in Warmoth Papers, SHC.

57. Wilbray J. Thompson to Warmoth, February 27, 1893, in Warmoth Papers, Reel 7, SHC.

58. *National Cyclopedia of American Biography*, XXIII, 307–308.

59. "James C. Murphy," *LP*, XL (1908), 371.

60. *Biographical and Historical Memoirs of Louisiana*, I, 329.

61. "J. B. Levert," *LP*, XLI (1908), 259. *Biographical and Historical Memoirs of Louisiana*, I, 545.

with Bush, established himself in the sugar, molasses, and cotton trade.

Four other prominent merchants played important roles in the early LSPA. One was David Calder (1830–1902), a Scotsman who fought for the Confederacy and who later conducted a large business with numerous sugar planters, eventually acquiring his own plantations. Another was William B. Schmidt (1823–1901). Schmidt, who was born in Germany, came to New Orleans in 1838 and formed a wholesale grocery business in 1845. Later he became involved in several banking and plantation enterprises and was closely associated with Richard Milliken in promoting the establishment of the Ear, Eye, Nose and Throat Hospital in New Orleans. Like Schmidt, John T. Moore (1838–1909) began as a wholesale grocer and later purchased several plantations. Finally, Leonce Martin Soniat (1841–1922), born in New Orleans, was educated at the University of Louisiana and the University of Virginia. After service in the Confederate army, Soniat developed a general merchandise business and later diversified into profitable sugar and lumbering activities.[62]

The leading planters and merchants within the LSPA were a highly educated, cosmopolitan, and politically sophisticated group of individuals who shared in the vision of a prosperous Louisiana sugar bowl. Because of economic realities, southern sugar planters like Duncan Kenner would not allow the hatred and ill feelings of the recent past to hinder future growth. Despite their politics, northern businessmen brought capital, energy, and expertise to the region. While planters from the old families of Louisiana would not as a rule socialize with carpetbaggers, they most certainly would learn from them and use them. And each faction needed the political power of the other if the interests of the sugar industry were to be fully protected in Washington.

LSPA Objectives: The Tariff

Immediately after organizing, the LSPA began to pursue well-defined goals. The association's most pressing objective was to thwart the

62. "David R. Calder," *LP*, XXX (1902), 322; "Mr. William B. Schmidt," *LP*, 26 (1901), 386; "Capt. John T. Moore," *LP*, XLII (1909), 259; *Who was Who in America* (1943 ed.), 1156.

efforts of those lobbying for a reduction in the import tariff rates on sugar. The high duties were viewed by many as a form of undemocratic taxation that was robbing the common man while filling the pockets of a small number of elite Louisiana planters. During the late 1870s influential economist David Wells asserted that the duties on sugar were twice as high as the average rates imposed on imported goods and that further increases should be opposed. Wells continued his criticism of the sugar duty, asserting that in contrast to a few privileged planters, "the general public, like the ass bowed down with heavy burdens, is alone complaisant."[63] Along with intellectuals like Wells and other citizens concerned with the fairness of existing legislation, large east coast refiners were also campaigning for lower rates on raw sugar.[64] The refiners were locked in a battle for markets with midwest grocers who were headquartered in Cincinnati. The grocers were selling a brown sugar of inferior quality to that of the refiners but at a much lower cost. This lower-grade product, made in Louisiana, was a key commodity in a lucrative trade that shipped sugar up the Mississippi and Ohio rivers and brought back goods normally not obtainable in the Deep South. Reducing the duties of lower-grade sugars while maintaining the rates on white, high-grade products—the aim of importers and refiners—would upset this regional trade and most certainly threaten the economic viability of the Louisiana sugar industry. Clearly, and for their own best interest, the LSPA leaders sought to ally themselves with the grocers but also reach an agreement with the refiners before a costly trade war erupted. It appears that the newly elected LSPA leadership quickly met with success in dealing with this potential business crisis. At the first official meeting of the Louisiana Sugar Planters' Association in January, 1878, Duncan Kenner announced that as the result of correspondence with the Cincinnati Grocer's Association, the grocers had altered their previous demand for a $0.01 horizontal tariff. They now favored a graduated tariff, one which placed a $0.025 tax on sugar below number twelve grade (dark brown in color), and

63. David A. Wells, *The Sugar Industry of the United States and the Tariff* (New York, 1878), 57; Robert G. Ingersoll, *A Review of the Sugar Question* (Washington, D.C., n.d.), 17; *The Government and the Sugar Trade* (New York, [1878]).

64. "Memorial of Importers, Refiners and Dealers in Sugar," [1878], in Joseph L. Brent Papers, Box 6, Louisiana State Museum, hereinafter cited as LSM.

$0.03 on sugar above that number. Compared with existing rates, these proposed duties were slightly higher on lowest grades and approximately $0.05 per pound lower on high quality products.[65] Kenner maintained that compromise and the union of planter and grocer interests had halted a movement that would eventually have eliminated the tariff altogether. Yet not all those attending the first meeting felt that the association represented the interests of all Louisiana planters. For example, Senator Goode "protested against the association sending the idea out to the world that the planters of Louisiana endorse the present tariff, which discriminates against the production of crude sugars, such as are nine-tenths of our sugar men. They favor a radical change in the tariff. He intimated that planters interested in the production of refined sugar had caused this action."[66] Quite correctly, Goode recognized that the LSPA was led by those planters who cultivated the largest acreage and who usually employed vacuum pans for the production of high-grade sugars. A differential duty favoring higher grades of sugar would clearly benefit this elite group.

During 1878 the LSPA sent a delegation to Washington to testify before the House Ways and Means Committee on proposed changes in the sugar tariff. As early as February of that year, John Dymond expressed confidence in future tariff legislation, and in March Kenner described Washington developments.

> The planters' committee took the ground that they were neither Republicans nor Democrats, but representatives of the whole people. There were two opposing interests at work; one was the grocers' interest and the other the refiners' interest. They commenced to talk with the planters' committee as if they desired a combination. The committee took a conciliatory course, and he [Kenner] thought the tariff bill as published will be exceedingly favorable to the sugar interest. He [Kenner] said that he thought the committee would return to Washington when the matter would come up before the House.[67]

65. "Dymond's Comparison of Sugar Duties" (MS in Brent Papers, Box 6, LSM).

66. *Louisiana Sugar Bowl* (New Iberia, La.), January 10, 1878.

67. Louis J. Bright, *New Orleans Price Current: Crop Year 1877–78* (New Orleans, 1878), 1.

When the negotiations ended, refiners and importers adopted the so-called Boston Plan, which met the approval of the LSPA; but as Goode had feared, the interests of the small growers were sacrificed.[68] The combined sugar interest requested that Congress levy high rates on refined sugar and on sugar produced in the better sugarhouses of Louisiana; the tariff on the low-grade product, however, would be levied at a rate of less than two cents per pound. For many of the members of the LSPA, the matter was settled favorably at the expense of marginal producers. Nevertheless, agitation over rates continued for the next decade. To complicate matters, the basis for judging the quality of sugar for revenue purposes was being questioned at this time, for the United States Treasury had become aware of an increasing number of sugar frauds. The bulk of sugar imported to the United States during the 1870s came from Cuba. As a result of the adoption of vacuum pans in that country, processed sugar that was high in saccharine content but also dark in color (due to the manner of processing or the misleading addition of coloring matter) entered the United States ports at a lower-than-legal duty. Customs agents used the so-called Dutch Color Standard rather than a polariscopic technique; thus, the United States government was being defrauded of large revenues. Clearly, some revisions to the existing tariff law remained to be worked out.

The tariff question quickly resurfaced again in October of 1879 when T. M. Cage, a Terrebonne Parish planter, read a short paper at the monthly LSPA meeting advocating the abandonment of the Dutch Color Standard and its replacement with polariscopic measurements.[69] Early in February, 1880, Kenner, Dymond, J. S. Wallis, Cage, and Taylor Beattie traveled to Washington to work against tariff reductions. One member of the House Ways and Means Committee was LSPA member Randall L. Gibson. The testimony centered around the rivalry between importers and grocers on one hand and refiners' interests on the other. Kenner testified for the Louisiana interests. He summarized the economic impact of the sugar industry upon the state and the need for protection because of high labor rates. Further,

68. New York *Daily Sugar Report*, August 11, 1882.
69. *Louisiana Sugar Bowl* (New Iberia, La.), February 5, 1880.

he maintained that whenever American duties were decreased in the past, Cuban authorities followed by raising export rates. Thus, the lowering of the sugar duty would not necessarily lead to a price reduction for American consumers. And Kenner, like many of his fellow Louisiana planters, advocated the use of the polariscope in grading the quality of sugar.[70] It is clear that Kenner, while looking out for LSPA interests, also compromised with the refining interests of New York and their leader, Theodore Havemeyer. In a letter written April 3, 1880, Havemeyer asked that Kenner exert influence upon Congressman Gibson so that a conflict between planters and refiners might be avoided.[71] Later, after the two major parties had reconciled, Dymond reported that "an interview had been had with the Committee on Ways and Means, and had been assured by that committee that no action inimical to the interests of Louisiana would be taken this Session."[72]

The Louisiana sugar interest, supported in Congress by Randall Lee Gibson, gained further representation by the appointment of Kenner to the 1882 United States Tariff Commission. The commission, which was established to investigate reductions in tariff rates, conducted numerous hearings from August 29 to October 16, 1882, receiving testimony from sugar refiners, wholesale grocers, and Louisiana planters.[73] However, the commission reports clearly reveal that Kenner controlled all discussions related to the sugar tariff. Both John Dymond and Harry McCall testified on behalf of the LSPA interests, citing the economic impact of the sugar industry on both the South and the Midwest and also the desire to maintain high wages and the employment of black laborers as major reasons for continued tariff protection. Dymond constructed an argument asserting that the welfare of the free black in Louisiana depended upon a viable sugar industry. Later he claimed: "The general good of our free institutions demands that our free laborers should be protected by a duty on those foreign sugars. . . . Shall we force our laborers

70. *Ibid.*, February 26, 1880.
71. Theodore A. Havemeyer to Duncan F. Kenner, April 3, 1880, in Dymond Family Papers, Folder 27, HNOC.
72. Louisiana Sugar Planters' Association, "Minutes of Monthly Meetings, 1877–1891," p. 20.
73. U.S. Tariff Commission, *Report* (2 vols., Washington, D.C., 1882).

into less productive industries that we may become missionaries to help the slave owner of Cuba, the coolie owners of Hawaii, the peon owner of Java and the heathen Chinese?"[74] The tariff commission recommended slightly reduced rates for the importation of sugar. While the rates on low-grade sugars were substantially reduced, the better grades, generally manufactured by the large producers of Louisiana, remained protected with high duties. This rate structure was the result of compromise between interests in the East and those in Louisiana, with Kenner playing the crucial role of mediator. An understanding had been reached by these two groups: in setting the duties for the highest grades of sugar, the refiners' wishes would prevail, while the LSPA would structure rates for darker sugars. Dymond expressed his feelings on the matter to Duncan Kenner in an October 30, 1882, letter in which he stated, "We have New York with us and Boston committed to the plan and now if you can carry the Tariff Commission I for one will appreciate the rare diplomatic power that gave us the victory."[75] Kenner's viewpoint prevailed and the LSPA had achieved its first objective and *raison d'etre*: adequate and certain protection for a marginally profitable industry.

LSPA Objectives: The Formation of the Sugar Exchange

Once the LSPA membership felt reasonably secure about future tariff legislation, the association began to consider plans for controlling the marketing of sugar in Louisiana. Sugar was normally sold on the levee in New Orleans, and the city-chartered Sugar Shed Company held a monopoly on commodity sales. In February, 1880, an LSPA committee reported on company abuses and mismanagement, concluding that the Sugar Shed Company's franchise was invalid and illegal. At an LSPA meeting held later that spring, Louis Bush cited existing problems in the storage, weighing, gauging, and handling of sugar on the levee. He asserted that the immediate task of the LSPA was "to bring suit against the Sugar Shed Company and destroy it."[76] Following Bush's recommendation, the LSPA raised funds necessary

74. *Opening Exercises of the Louisiana Sugar Exchange of New Orleans, 3 June 1884* (New Orleans, 1884), 35. See also Sitterson, *Sugar Country*, 325.

75. John Dymond to Duncan F. Kenner, October 30, 1882, in Brent Papers, Box 6, LSM.

76. *Louisiana Sugar Bowl* (New Iberia, La.), February 19, 1880, and April 22, 1880.

for legal action against the Sugar Shed Company and the city of New Orleans. Concurrently, the membership sent a petition to the city attorney asking that the city council ratify a charter establishing a new sugar exchange.[77]

In March, 1883, the charter was granted to the Louisiana Sugar Exchange, and $52,000 was raised for the construction of a building.[78] The purpose of the exchange was to coordinate the marketing activities of planters and brokers. Its members had the power to set rules and regulations for the sale and delivery of sugar, to penalize those unable to fulfill contracts, and to set standards of weighing and gauging.

Twelve directors controlled the exchange: six were planters and sugar factors, and six were sugar brokers and dealers. Edward J. Gay, an active LSPA member, was the first president of the exchange, and John Dymond the vice-president. Members of the LSPA who held seats on the exchange included D. R. Calder, L. C. Keever, John T. Moore, Jr., Adam Thomson, and R. H. Yale. Several other prominent members of the LSPA were officers of the exchange, including Louis Bush, Hanson Kelly, Richard Milliken, and Bradish Johnson.

Dymond felt that if the Louisiana sugar industry was to compete against European beet sugar, an LSPA-controlled exchange was necessary. He asserted:

> Then why do we change? All the world has changed and our turn has come. The western merchant today tells us he does not want our sugar unless it is white. . . . What does all this mean? It means that in sugar we are now competing with all the world, and that the sugar business contains little or no margin, and that to succeed in it every device suggested by the modern economy must be availed of. Sugars must be bought and sold quickly and correctly, and at a minimum of expense; disputes must be quickly settled by arbitration; abuses must be corrected by authority; new good methods must be brought out and old bad ones rejected; and modern experience shows that this can all be best done by an Exchange.[79]

77. A. Bouchereau, *Statement of the Sugar and Rice Crops Made in Louisiana in 1879–1880,* xvi.

78. *Charter, Constitution, By-Laws and Rules of the Louisiana Sugar and Rice Exchange of New Orleans, Louisiana Incorporated 6 March 1883* (New Orleans, [1883]).

79. *Opening Exercises of the Louisiana Sugar Exchange,* 33.

With the organizational goals of tariff legislation and market control realized, Dymond and the LSPA increasingly emphasized the role of science and new technology in stemming the tide of international competition. In his speeches to the LSPA, Dymond had asserted that every device had to be tried. While politics and marketing were very much a part of the nineteenth-century sugar planters' world, science and science-based technology were not. The injection of scientific and technical knowledge into the Louisiana sugar industry proved to be the major challenge of LSPA leaders during the 1880s. This process of technological modernization was facilitated by the same strategy that brought the planters together in 1877—organization.

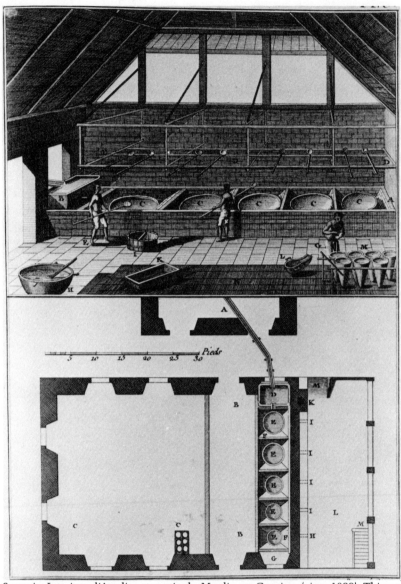

Sucrerie, Interieur d'Attelier et partie du Moulin ou Goutiere (circa 1800). This en-
graving shows the interior of a late-eighteenth-century sugarhouse equipped with
open kettles. Juice expressed from the roller mill was clarified and crystallized in
these vessels.
Special Collections, Hill Memorial Library, Louisiana State University

Leeds Foundry, New Orleans. Established in 1825, Leeds Foundry was an important institution for the introduction and fabrication of advanced design process equipment.
The Historic New Orleans Collection

Sugar hogsheads on the levee, New Orleans. During the nineteenth century, most of the sugar made in Louisiana was marketed in the Midwest.
The Historic New Orleans Collection

John Dymond (1836–1921). Dymond, a Plaquemines Parish planter and sugar manufacturer, played a prominent role in the establishment of the Louisiana Sugar Planters' Association, the Louisiana Sugar Experiment Station, and the Audubon Sugar
School.
The Historic New Orleans Collection

Evan Hall Plantation, 1888. Located in Ascension Parish, Evan Hall was owned by the McCall family and was one of the largest central factories of the day.
Special Collections, Hill Memorial Library, Louisiana State University

United States Department of Agriculture chemists conducting sugar trials, fall, 1885. Chief chemist Harvey Wiley and his assistants played a major role in stimulating scientific and technical activity in the late-nineteenth-century Louisiana sugar industry. From left to right: W. C. Parkinson, Magnus Swenson, Clifford Richardson, John Trinkle, Harvey W. Wiley, John Dugan, unidentified, Guilford Spencer.
Library of Congress

5

Responding to the Competition

The success of the Louisiana Sugar Planters' Association in nego-
tiating a favorable tariff for the elite planters of Louisiana was in
large part due to the political experience of its leadership. However,
John Dymond and others felt that a protective tariff simply was not
enough to stem the tide of foreign competition. In their opinion, science and advanced technology—the basis of success for the European beet-sugar industry—would have to be transplanted and systematically applied to cane cultivation and manufacture if the sugar
bowl were to truly prosper. However, LSPA leaders were not professional scientists and engineers, and to inject new methods into a
tradition-bound industry would be a formidable task. Ultimately
this new strategy reoriented the basic objectives of the association.
Once its leadership discovered that existing organizations were of
limited usefulness, the LSPA became a patron of scientific, technical, and educational institutions during the 1880s.

LSPA Objectives: Science

From the inception of the association it was clear that LSPA leaders
recognized the necessity of developing alliances with scientists.
In January, 1878, Joseph L. Brent read to the planters a paper entitled "The Necessity of Science to a Just Development on the Sugar
Culture." Brent recognized that the Louisiana sugar industry was a

"creature of government protection; and even this protection will prove insufficient to sustain it unless it be aided by all the resources of modern science."[1] He understood that Louisiana's short growing season and high wage rates placed the industry in a noncompetitive situation relative to Caribbean producers.

Despite the dire straits in which the planters found themselves, the majority seemed to be wedded to old rule-of-thumb ways and unsystematic practices. In many cases, there was no rational reason why certain planters did what they did in the fields and in the sugarhouse. To complicate matters, Louisiana planters employed a variety of methods for planting and cultivating cane and manufacturing sugar. Brent asserted that "these divergences of opinion manifest that somebody is wrong, for all cannot be right." He desired fixed, exact knowledge of what should and should not be done in growing and making sugar. However, he realized that this knowledge could not come from the work of individual planters. He suggested that a proposed comparative study on milling "must be conducted as an experiment alone, under the supervision and control of science, and under conditions which, in the chemical laboratories, give accurate and incontrovertible results."[2]

Brent felt that government-sponsored research would resolve the numerous controversies among planters. This idea originated not only from his examination of successful agricultural research conducted at Cornell and the Agricultural College of Massachusetts—state supported agricultural experiment stations—but more specifically from his observations of European institutions, "where governments caused every study and experiment to be made exhaustively by science, which has thus been enabled to create a great industry, worth to Europe thousands of millions of dollars." Brent thought that the easy answer was to link the planters' needs to the state university at Baton Rouge. He moved that a committee of seven be appointed to consult with professors David French Boyd (the president of Louisiana State University) and Richard Sears McCulloch (professor of agricultural chemistry) on the necessary apparatuses to conduct experiments. In the LSPA minutes it was resolved "that after

1. Louis J. Bright, *New Orleans Price Current: Crop Year 1877–78* (New Orleans, 1878), xxxvii.

2. *Ibid.*, xxxviii.

such conference the said committee shall [have] authority to apply to the legislature for an appropriation sufficient to provide the State University and Agricultural College with such apparatus and material as may enable the proper . . . development of the sugar culture in this state to be sought by a series of exact experiments conducted by the said college under conditions insuring their correctness and reliability."[3]

McCulloch had had a checkered career after the publication of his controversial but important study on Louisiana sugar technology during the 1840s.[4] After a brief appointment as melter and refiner at the United States Mint at Philadelphia, in 1849 he was appointed professor of natural philosophy at Princeton. He remained at Princeton until 1854, when he was selected professor of natural and experimental philosophy and chemistry at Columbia College. In 1863 he resigned from Columbia to serve with the Confederate army. After his release from a Federal prison in Richmond in 1866, he was hired by Robert E. Lee as professor of natural philosophy at Washington College. His 1871 lecture notes on the mechanical theory of heat were revised and published in 1876 as *A Treatise on the Mechanical Theory of Heat and Its Applications to the Steam Engine*. In 1877, after an unsuccessful bid to become professor of chemistry at the new Johns Hopkins University, he returned to Louisiana, where his 1848 study on sugar manufacturing was still read by several important members of the newly founded LSPA.

While we have no record of the outcome of LSPA meetings with the university, the membership apparently remained hopeful that McCulloch would assist them. In May, 1878, the association passed a resolution stating that because of the numerous innovations and rapid growth of the European beet-sugar industry, "a scientific man should visit Europe and by personal inspection examine the various late improvements made in this industry and report the same to the La. Sugar Planters' Association. . . . Therefore be it resolved, that in order to facilitate the carrying out of this purpose the gov. of Louisiana be requested to appoint Prof. Richard S. McCulloch honorary

3. *Ibid.*, xxxix; Louisiana Sugar Planters' Association, "Minutes of Monthly Meetings, 1877–1891," January 3, 1878, p. 12.

4. Thomas, "Professor McCulloch of Princeton, Columbia and Points South," 17–29.

Commissioner for the state to the Paris Exposition." However, there is no record of McCulloch's trip to Paris. The professor may well have been opposed to accepting additional responsibilities in assisting the sugar planters. In a letter written to President Boyd on August 10, 1878, McCulloch stated: "As for the Agricultural Department, I have not regarded it as under my charge, except in so far as 'Agricultural Chemistry' is concerned; nor do I desire further responsibility—for it. Indeed, the farm seems likely to be an 'Elephant' which the supervision will manage *as they please.*"[5]

Yet McCulloch did not close the door to all LSPA inquiries. In June of 1879 he apparently considered revising his 1848 treatise on sugar. And in late 1879 the LSPA discussed a plan in which he would visit a number of Louisiana plantations to study the clarification and defecation processes.[6]

McCulloch's association with the LSPA continued through the mid-1880s. He published an essay under the auspices of the LSPA in 1886 entitled *On the Present State of Knowledge of the Chemistry and Physiology of the Sugar Cane and of Saccharine Substances.* This rambling discourse attempted to explain, in simple terms, the nature of saccharine substances and their chemical transformations. Natural products such as gums, albuminoids, starch, cellulose, and glucose were characterized, and fermentation processes were briefly discussed. Since the essay contained little new information and since McCulloch had borrowed heavily from his 1848 treatise, this work was probably a disappointment to the membership of the LSPA.[7] Just as he did in his 1848 study, McCulloch emphasized the studies of French scientists like Vauquelin and Peligot, and he quoted a long passage from the recently published diary of Valcour Aime. But the LSPA leaders of the 1880s were thirsting for recent studies derived

5. Louisiana Sugar Planters' Association, "Minutes of Monthly Meetings, 1877–1891," May 2, 1878, p. 20; R. S. McCulloch to Thomas D. Boyd, August 10, 1878, in Walter L. Fleming Papers, Alphabetical File 890, Special Collections, Hill Memorial Library, LSU.

6. *Louisiana Sugar Bowl* (New Iberia, La.), June 12, 1879; Louisiana Sugar Planters' Association, "Minutes of Monthly Meetings, 1877–1891," November 6, 1879, p. 35.

7. R. S. McCulloch, *On the Present State of Knowledge of the Chemistry and Physiology of the Sugar Cane and of Saccharine Substances* (New Orleans, 1886), 11–12.

from the German and French beet-sugar industries, not the information that had been of interest to Louisiana sugar planters during the 1840s.

McCulloch's apparent lack of cooperation or his inability to keep abreast of recent developments may not have been the only stumbling block in the LSPA's design to forge a strong alliance with local academic institutions. William Preston Johnston, president of Louisiana State University for a short time during the 1880s, was willing to establish a sugar laboratory on the campus, but perhaps his financial demands were more than a new planters' organization could meet. Johnston, in an 1881 letter to the editor of the *Louisiana Sugar Bowl*, stated: "One distinct function of this university will be to aid in the development of the sugar industry. . . . I am seeking for means to erect a sugar laboratory. . . . Certainly, either from State aid, or private generosity, we might and should receive the small amount necessary to erect a laboratory. $10,000 would be sufficient, and the possible benefits are enormous."[8] Johnston made no promises that the sugar laboratory would attempt to be responsive to the immediate and future needs of the Louisiana sugar interest. Thus, Johnston's request was perhaps considered unreasonable by this group of planters confronted with both high interest rates and labor shortages. The planters wanted a trade laboratory and an experimental farm that would focus solely on agricultural and manufacturing problems, yet the state university appeared hesitant to accept such responsibilities without a large sum of cash.

The failure of the LSPA to secure the permanent assistance of university scientists during the early 1880s may have determined, at least in part, the organization's future strategy. Rather than depend upon scientists associated with independent institutions, the LSPA decided to create scientific institutions totally responsible to the membership.

LSPA Objectives: New Technology

In addition to their attempts to develop a scientific basis for their industry, the LSPA began to investigate new methods of sugar manufacture. European beet-sugar manufacturers were perceived to be a

8. *Louisiana Sugar Bowl* (New Iberia, La.), May 6, 1881.

growing threat, and the circular issued on November 29, 1877, by LSPA founders stated: "Again, we find that while we have been standing still in cane cultivation and sugar production during the military occupation of the State, a period of fifteen years, the world has moved on, and everywhere else sugars are produced in greater abundance and at less cost than with us at present. A standard European authority recently quotes beet juice weighing 10-½ degrees Beaumé, and as pure as our cane juice: and Europe is now producing, from beets, about two-thirds as much sugar as is produced from cane in the rest of the world." As the editor of the *Louisiana Sugar Bowl* remarked, "Nothing succeeds like success, and of this no stronger proof could be given than the extraordinary and rapid increase of the sugar production in Austria."[9] The LSPA leaders had correctly perceived that they were a part of a dynamic international market, and their dilemma was how to respond to it.

It was hoped that the application of European methods and machinery to Louisiana sugarcane processing would enable them to remain competitive. In early 1879, Rudolph Sieg, a frequent contributor to the *Louisiana Sugar Bowl*, wrote: "A lively competition must now be expected, and our planters themselves will see the necessity of meeting the attack from this new quarter with the same weapons which are employed by their rivals, namely a more rational and scientific cultivation,. . . and they will also be compelled to cast aside much of their . . . machinery."[10]

During 1879 and 1880, John Dymond constantly reiterated Europe's menacing competitive advantage at the monthly LSPA meetings. As chairman of the executive committee, Dymond normally commented on European developments at the beginning of each meeting, frequently reading translations of important articles from the French *Journal des Fabricants de Sucre*. Dymond concluded that while Louisiana planters could remain competitive in agricultural developments, "we must look to the manipulation of the cane after it has left the hands of the producer, and while it is in the hands of the manufacturer." Dymond became a strong advocate of the systematic investigation of the diffusion process. At a monthly meeting

9. Bright, *New Orleans Price Current: Crop Year 1877–78*, xxxiv; *Louisiana Sugar Bowl* (New Iberia, La.), November 30, 1879.

10. *Louisiana Sugar Bowl* (New Iberia, La.), September 4, 1879.

in April, 1879, he stated: "The dearly learned lessons of the past bid fair to so improve diffusion, as applied to sugar cane, that the cane planters adopting it shall so increase their yield and so diminish their expenses that they will be able to hold their own against this European giant, grown suddenly so strong that he now seems to crush the cane-producing world out of existence." Dymond felt that the widespread acceptance of diffusion in Germany and Austria, as well as the fact that French manufacturers were replacing their presses with diffusion vessels, was sufficient evidence of the possible applicability of the process in Louisiana.[11]

The LSPA's initial attempts to improve manufacturing efficiency included a comparison of process data gathered from a number of sugarhouses. Joseph L. Brent and other LSPA leaders felt that a comparative examination of the various mills and milling techniques employed in Louisiana would logically lead to proper conclusions concerning the most efficient practices. In March of 1879 Brent proposed that the LSPA's secretary collect process data from all planters in Louisiana who used weighing scales in their sugarhouses.[12] He assumed that information concerning the percentage of juice from cane, as well as the percentages of dry sugar and molasses manufactured, would clearly indicate both the best and worst manufacturing apparatuses and techniques. However, the study was disappointing and inconclusive, since the cooperating planters used different methods of calculation to report their results.

Similar studies were sponsored by the LSPA in September, 1880. To avoid the shortcomings experienced during the first evaluation, the association sent a circular to planters outlining the proper method for determining the weight of cane employed, the number of gallons of juice expressed from the mill, and the weight of byproduct bagasse. James F. Giffen, the secretary of the LSPA, was given the task of compiling the statistical production data and crop reports from the participating Louisiana planters. This information was normally summarized annually and presented to the association at its March meeting. Information obtained from this survey included the

11. *Ibid.*, July 21, 1879, and September 16, 1880.
12. Louisiana Sugar Planters' Association, "Minutes of Monthly Meetings, 1877–1891," February 6, 1879, p. 26.

type of fertilizers employed and the method of application, as well as the condition of the stubble, seed cane, and acreage.[13]

The LSPA also appointed committees to examine new processes and the improvement of machinery already in use. For example, a committee of Kenner, Wilkinson, and Yale studied extraction efficiencies of various mills and the purity of expressed juice. The inconclusive results obtained from a set of poorly conducted experiments during 1881 indicated that, for future studies, the Louisiana sugar interests would have to obtain scientific and technical expertise outside the planter community.[14]

Since its own initial attempts to conduct engineering studies had ended inconclusively, in the 1880s the LSPA increasingly relied on the expertise and creativity of the foundrymen to improve manufacturing machinery. The leadership of the LSPA quickly recognized the inability of its planter-merchant members to provide the technical expertise necessary to meet the challenges posed by foreign competition.

In an effort to critically examine new processes and machinery, the LSPA began to ally itself to the foundry community in 1879. During its first year, 1878, all but two papers presented to the association had been given by Louisiana planters. However, between 1879 and 1880 discussions focused on methods of juice extraction, and a number of foundrymen, including Archibald Mitchell of Leeds and Company and A. W. J. Mason, presented papers at these meetings. Different designs of mills were discussed, as well as the problem of determining the proper mill speeds for maximum extraction efficiency. Of the twenty-six topics presented at LSPA monthly meetings during 1879–1880, sixteen dealt with mechanical methods or the diffusion process.[15] In this institutional setting, the planter-foundryman relationship that had first emerged during the antebellum period was strengthened. Although foundry personnel were never elected to positions of power within the association, the LSPA's

13. *Louisiana Sugar Bowl* (New Iberia, La.), September 16, 1880, and June 16, 1881.

14. *Louisiana Sugar Bowl* (New Iberia, La.), July 22, 1880, and July 29, 1880.

15. Louisiana Sugar Planters' Association, "Minutes of Monthly Meetings, 1877–1891," 26–56.

emphasis upon the introduction of new technology during the 1880s was clearly the result of the participation of practical men in this elitist organization. Foundrymen viewed the association's meetings as an opportunity to display their inventions and improved process apparatuses. As a consequence, New Orleans foundries thrived during this period of rapid technological change.

While the significance of the machine tool industry to the development of nineteenth-century technology has been emphasized by Nathan Rosenberg in his *Technology and Economic Growth*, historians have totally neglected the central institution of this trade—the foundry.[16] As Rosenberg has so capably pointed out, the machine tool industry served as a focal point for the transmission of new technology to a variety of manufacturing activities, including the sugar industry. Yet the foundry was more than a source of innovations, it was also a training ground for a host of apprentices who would become skilled practical engineers. What were these places like, what kinds of inventions were developed, and who were the apprentices and journeymen who spent their lives working there?

A number of foundries, having survived the disruptive effects of the Civil War, had flourished during Reconstruction. Leeds and Company, founded in 1825, was the most important of these local businesses, for it employed two innovative practical engineers and trained several other practical engineer-chemists. During the early 1880s the company advertised itself as manufacturers of "Vertical and Horizontal Steam Engines, Sugar Mills, Vacuum Pans, Clarifiers, Filters, Bone-black Revivifiers, Boilers of every description, Bagasse Furnaces, Sugar Kettles, Furnace Parts, Grate Bars, Steam and Horse-Power Draining Machines, Cotton Compressors and Gin Gearing."[17]

The Leeds foundry consisted of a pattern shop, molding floor, machine or finishing shop, blacksmith shop, and boiler shop. Assisted by hoisting machinery, workmen in the pattern shop employed various types of wood saws—circular, rip, crosscut, suspended, and band—along with lathes in the manufacture of custom patterns. A watertight cast-iron pit, twelve feet in diameter and twenty feet

16. Nathan Rosenberg, *Technology and Economic Growth* (New York, 1972), 87–107.

17. John S. McDonald, *Automatic Hydraulic Pressure Regulator for Sugar Mills* (New Orleans, 1884), inside cover.

deep, was located in the middle of the molding floor, an area de-scribed as "a vast theater of laboring activity."[18] The castings were made there, and then subsequently moved by hoisting cranes to a finishing department. Large lathes, as well as planing and boring mills, shaped the castings to fit the design specifications. The black-smith shop contained a number of steam hammers, while in the boiler shop, machinery was used to bend, punch, and rivet metal.

Leeds and Company employed several innovators who specialized in the design of large-scale apparatuses for the sugar industry. John McDonald began as an apprentice for Leeds in 1846.[19] After studying feeding problems of the rigid roller mill, he patented a hydraulic pressure regulator in 1872. Previously, the compressibility of wood served as a compensation for variations in mill feed. Mill pressure with rapid rollers was variable, often alternating between a danger-ous maximum with a heavy feed to a light pressure with a light feed. At high pressure, mill rollers would often break, while poor juice ex-traction resulted from low pressures. McDonald's pressure regulator applied a high but safe and constant pressure to the rollers. After a strong endorsement by Ascension Parish planters, including Duncan Kenner, the device was employed on many Louisiana sugar planta-tions, including those of Richard Milliken, H. P. Kernochan, Bradish Johnson, and T. S. Wilkinson.

Archibald Mitchell was the general superintendent of Leeds dur-ing the 1870s and 1880s. Born in Scotland in 1819, Mitchell came as a young man to New Orleans in 1840. There he was employed as a journeyman mechanic at Leeds, and later as chief superintendent of the foundry.[20] He patented several devices to improve the efficiency of roller mills and the transport of sugarcane during processing. Mitchell was also a frequent contributor to the monthly LSPA meet-ings. His practical mechanical experience was a welcomed, indeed necessary, addition to the association's membership of merchants and planters.

18. *Louisiana Sugar Bowl* (New Iberia, La.)., February 13, 1879, December 19, 1879, and February 13, 1879.

19. "McDonald's Hydraulics," *LP*, XL (1908), 57.

20. Obituary of Archibald Mitchell, in Lucille Mouton Griffin Collection, Scrap-book, Vol. 1, Southwestern Archives and Manuscripts Collection, University of South-western Louisiana, hereinafter cited as USL.

The Leeds foundry also provided the opportunity for several individuals to begin their careers in sugar technology. Lezin Becnel began there as a machinist's apprentice in 1878.[21] Subsequently, he entered the drafting and design departments. During the 1880s he managed the McCall brothers' sugarhouses in Ascension Parish and was responsible for both the chemical and mechanical departments of the central factory. Later he became a consulting chemical engineer and was a charter member of the American Institute of Chemical Engineers in 1908.

Alexander Mouton, the son of an antebellum governor of Louisiana, followed a similar career path.[22] As a young man, Mouton left Lafayette, Louisiana, for New Orleans, where Mitchell and others at the Leeds foundry taught him basic machine tool skills, including milling, boring, and operating a lathe. After serving an apprenticeship of more than three years, Mouton worked for the Pennsylvania Railroad, the Louisiana Western Railroad, and the United States Mint before embarking on a long career as a sugar engineer in Cuba and South America. Both Mouton's and Becnel's educational experiences occurred during a time when university education in engineering was the exception rather than the rule. Because practical men were often taught at the leading foundries, the machine tool trade was not only a source for the diffusion of new technology, but also a place to train technical experts.

Between 1880 and 1900, a number of immigrants played important roles in the New Orleans foundry community. Burchard Thoens, who was born and educated in Germany, was hired in 1877 as a machinist at Leeds, where he was later promoted to draftsman. Eventually he became the superintendent of a competing foundry, where he specialized in improvements in sugar mills, crushers, and vacuum pans. Another prominent foundryman involved in LSPA activities was Ernest N. Loeb.[23] An Alsatian by birth, Loeb became a foreman in the firm of James D. Edwards and later entered into a partnership

21. "Trade Notes," *LP*, XXIV (1900), 84.

22. Alexander Mouton, "Autobiography" [March, 1904?] (MS in Griffin Collection, USL), 233–42; Archibald Mitchell to Col. Adams, May 4, 1880, in Griffin Collection, USL.

23. "Mr. Burchard Thoens," *LP*, XXX (1903), 19. "Ernest M. Loeb," *LP*, XLI (1908), 83.

with Leon Haubtman, another practical machinist. The firm of Edwards and Haubtman grew rapidly during the 1880s, supplying planters with such devices as Hepworth Centrifugal Machines, double- and triple-effect evaporators and Cooks' Hot Blast Bagasse Burner.[24]

While the Louisiana Sugar Planters' Association had several successes between 1878 and the early 1880s, it had its failures as well. With the exception of the foundry group, the planters lacked the scientific and technical expertise necessary to continually present worthwhile papers and conduct systematic investigations. At times, the meetings were merely rambling and diffuse discussions.[25] Opponents of the association felt that it was all talk and no action. In addition, some planters felt that their individualism would be restrained by association membership.

The LSPA constantly made efforts to increase membership. In an 1878 letter, Kenner wrote: "Many persons think that all will go along well enough without their aid, and never attend one meeting, though their sympathies are with us. To such we say you do not appreciate the enthusiasm generated by members. At every meeting every new face that appears encourages those who are devoting themselves to the work, and it is certainly not asking much of every planter to so arrange his visits to the city that when he does come, he can attend a monthly meetin [sic]."[26] The editor of the *Louisiana Sugar Bowl* also defended the association. "If the association *has* done much talking, some of it not suiting the ideas of everybody, it was about all they could do, with their scant membership. You surely did not expect them (those few) to buy for you such information as you want? . . . Yet, such would seem to be your meaning, as long as you stand aloof and criticize without contributing one mite. . . . If they are wrong, go in and help to put them straight."[27]

The LSPA, always cognizant of the importance of government assistance to the success of their industry, recognized that assistance

---

24. Edwards and Haubtman, *Illustrated Catalog of Edwards and Haubtman* (New Orleans, 1889), 198–203.

25. *Louisiana Sugar Bowl* (New Iberia, La.), April 21, 1881.

26. Reprinted in U.S. Treasury Department, *U.S. vs. 712 Bags of Sugar Imported in the Mississippi*, 31.

27. *Louisiana Sugar Bowl* (New Iberia, La.), July 25, 1881.

from the United States Department of Agriculture was crucial for obtaining the necessary scientific expertise. During the early 1880s, the LSPA began to make overtures to the USDA in hopes that the federal government would support its objectives. They found a willing ally in USDA Commissioner William G. LeDuc, who had expressed an interest in the activities of Louisiana sugar planters as early as 1877. The LSPA attempted to establish a relationship with the commissioner by sending a delegation to visit him in Washington. In 1881 the association also passed a resolution thanking the commissioner "for his uniform courtesy and kindness towards the members of the Association, and for the general interest manifested by him in the interest of sugar cane cultivated in Louisiana."[28] Although the USDA accomplished little in Louisiana under LeDuc's leadership, the groundwork was laid for the strong relationship that later developed between USDA leaders and LSPA members. Indeed, the presence of the United States Department of Agriculture in Louisiana during the 1880s was an important driving force behind the late-nineteenth-century modernization of the state's sugar industry.

28. *Ibid.*, March 17, 1881.

6

Planters and the United States Department of Agriculture: Compromise and Accommodation

By the early 1880s the Louisiana Sugar Planters' Association had developed an effective organizational policy to deal with federal tariff legislation and had established the Louisiana Sugar Exchange to standardize marketing procedures. In addition, LSPA committees began to investigate and evaluate new designs in process technology; association leaders had recognized early on that the Louisiana sugar industry's greatest weakness was a lack of technical and scientific expertise. One strategy advanced to meet this need was the recruitment of local foundrymen and practical engineers into the LSPA. However, it was clear to some, like John Dymond, that the sugar industry was in the midst of a technological revolution based primarily on European ideas and that these new engineering techniques and chemical methods involved the services of university-trained scientists and not simply men educated through practical experience. Yet Louisiana had few chemists and engineers with formal training. To remedy this situation the LSPA courted the United States Department of Agriculture. As the association quickly discovered, the USDA was a willing partner. The USDA's cooperation with the LSPA during the 1880s was the result of the agency's commitment to a program of national self-sufficiency in the production of sugar and the recognition that the program could succeed only if it had the political support of Louisiana congressmen and senators who were members of the agency's committees. Therefore, the USDA's institutional strength was in large part the consequence of its ability to extend its influence outside Washington. The support of the LSPA was crucial to obtaining appropriations and maintaining a research and outreach program.

Between 1884 and 1888 this federal agency not only supplied the state with a large corps of formally trained research scientists, but also facilitated the transfer of European developments in the chemical control of manufacturing processes and new plant apparatuses. As a result of USDA interest in the Louisiana sugar industry, the widespread use of systematic chemical analysis to monitor large-scale processes became prevalent, and new clarification, filtration, and evaporation equipment was introduced.

While the USDA's presence in Louisiana had a marked influence upon the modernization of the state's sugar industry, the LSPA in turn significantly affected USDA policy and objectives. Since the USDA chemists, led by chief chemist Harvey Wiley, employed foreign machinery in their trials and advocated the diffusion process, their efforts drew the criticism of those foundrymen and planters tied to traditional milling techniques and rule-of-thumb methods. But even though early USDA investigations were inconclusive, the LSPA continued to openly support the agency, both politically and financially. As a consequence of the strong institutional alliance between the USDA and the LSPA, Wiley was able to survive the attacks on his experimental programs at this crucial juncture in his career.[1] Indeed, the USDA's Bureau of Chemistry grew in both size and power under Wiley's leadership during the 1880s.

The USDA and Sorghum Sugar Fever: Early Post-Reconstruction Attempts to Assist the Louisiana Sugar Industry

With the inauguration of Rutherford B. Hayes as the nineteenth president of the United States, the painful period of Reconstruction

1. Gladys L. Baker, Wayne D. Rasmussen, Vivian Wiser, and Jane M. Porter, *Century of Service: The First 100 Years of the United States Department of Agriculture* (Washington, D.C., 1963), 19–25; Gustavus A. Weber, *The Bureau of Chemistry and Soils: Its History, Activities and Organization* (Baltimore, 1928); A. Hunter Dupree, *Science in the Federal Government: A History of Policies and Activities to 1940*

ended in Louisiana. In a sense, Hayes owed his office to political negotiations centered on the contested 1876 Louisiana election returns. It is no surprise, then, that the USDA displayed remarkable
sensitivity to the problems of the Louisiana sugar planter during the
first months of the Hayes administration.[2]

Between 1865 and 1876 the commissioner of agriculture rarely
mentioned Louisiana in his reports; however, between 1877 and
1878 the state's needs became a new priority for the USDA in Washington. Within months of entering office, William G. LeDuc, a Min-
nesotan appointed commissioner of agriculture by President Hayes,
considered government assistance for Louisiana. LeDuc was interested in Louisiana not only for political reasons, but also because he
felt that Louisiana was integral in his plan for national economic recovery. He asserted that the money spent on imported sugar in the
United States should be redirected to support sugar growers and
manufacturers.

> The sugar-interest of the country reaches every cupboard in our broad
> land, and is intimately connected with every branch of the Inter-State
> trade and commerce of the republic; and if the manufacture of sugar be
> encouraged and developed to the extent of supplying our home demand
> with home grown sugars, importations will necessarily cease, and the per-
> petual flow of American gold to countries with which we have com-
> paratively no trade will be arrested, specie resumption will be assured, . . .
> and the material prosperity of the country will advance with renewed en-
> ergy and power under the changed condition of production of this single
> article of universal consumption.

In his 1877 *Report*, LeDuc emphasized the need for repairing the neglected levee system in Louisiana and asserted that only the "general
government is adequate to protect this wide expanse of fertile territory [Louisiana], and give confidence to capital and labor to again oc-

---

(New York, 1964), 149–83; Oscar E. Anderson, *The Health of a Nation: Harvey Wiley
and the Fight for Pure Food* (Chicago, 1958); William Lloyd Fox, "Harvey W. Wiley's
Search for American Sugar Self-Sufficiency," *Agricultural History*, LIV (1980), 516–26;
Harvey W. Wiley, *An Autobiography* (Indianapolis, 1930).

2. C. Vann Woodward, *Origins of the New South, 1877–1913* (Baton Rouge, 1951),
23–50; Benjamin F. Rogers, Jr., "The United States Department of Agriculture and the
South, 1862–1880," *Studies* (Florida State University), X (1953), 71–80.

cupy and cultivate it." LeDuc continued: "It is a *national* work, for a *national* purpose, and, as it seems to me, a *national* duty at this time to take in hand and push to a speedy conclusion the re-establishment of the broken levees, and the making of such provision for their maintenance as shall permanently secure the valuable industries that will immediately reoccupy the lands now subject to overflow."[3]

The commissioner felt that the Louisiana economy was integrally tied to that of the nation, and that the reestablishment of a profitable sugar industry would stimulate the growth of related industries, both in the Midwest and in the East. LeDuc perceived that if the Louisiana sugar industry was restored to its antebellum prominence, its increased volume of products would reduce the quantities of imported sugar in the United States. Hence, a revitalized Louisiana sugar bowl would help alleviate the nation's balance of payments problems.

In an effort to better understand conditions in Louisiana, LeDuc sent a questionnaire to numerous planters in 1877 asking them to advise him on how his department could promote the sugar interest. In their replies, the planters emphasized the poor condition of levees, the need for an improved seed cane, the labor shortage, and a lack of scientific experts in Louisiana to conduct soil and fertilizer analyses. One planter, H. von Puhl of East Baton Rouge, recommended the development of new process methods and apparatuses and proposed "the immediate establishment, by State or otherwise, of an agricultural station or experimental farm."[4]

Although LeDuc had the best of intentions, he effected only small changes that for the most part went unnoticed. Through his assistant chemist, he sent new varieties of cane to several planters; later

3. U.S. Department of Agriculture, *Report of the Commissioner of Agriculture for the Year 1877*, 7; On LeDuc, see William G. LeDuc, *Recollections of a Civil War Quartermaster: The Autobiography of William G. LeDuc* (St. Paul, Minn., 1963); Charles H. Greathouse, *Historical Sketch of the U.S. Department of Agriculture: Its Objects and Present Organization* (Washington, D.C., 1898); *Extracts of a Few Letters and Notices Received at the Department of Agriculture During the Administration of Gen. Wm. G. LeDuc* (Washington, D.C., 1881); *National Cyclopedia of American Biography*, XXIV, 330–31.

4. U.S. Department of Agriculture, *Report of the Commissioner of Agriculture for the Year 1877*, 23, 41–44.

he met with an LSPA delegation, comprised of Kenner, Dymond, and others, to discuss the planters' problems. Yet his initial enthusiasm was dampened by the deterioration of the relationship between the Hayes administration and southern congressmen, and by his increased comprehension of the economic obstacles facing Louisiana planters. By late 1878 the Hayes administration had abandoned its earlier policy of reconciliation with the South. Southern congressmen had failed to garner the support necessary to elect James Garfield Speaker of the House, and the union of southern agrarian interests with those of western farmers alienated eastern Republicans in Washington. In addition, LeDuc realized that labor problems, a shortage of capital, and the need for large-scale improvements in the levee system would take years to overcome, and Louisiana's current resources were simply inadequate to supply all the sugar needed in the United States. In the opinion of the commissioner of agriculture, the nation's self-sufficiency depended upon the commercial development of additional production regions and the successful cultivation of alternative plant sources for sugar.

Initially, LeDuc had hoped that the sugar beet could be cultivated by farmers in those regions where its growth was economically feasible. To that end, he sent his chief chemist, William McMurtrie, to France in 1878 to examine and report on European agricultural and manufacturing practices. McMurtrie's *Report on the Culture of the Sugar Beet and the Manufacture therefrom in France and the United States* described the varieties of beets grown in France, soil preparations and proper planting practices, harvesting methods and manufacturing processes.[5] Most significantly, McMurtrie discussed the optimum meteorological conditions for beet-root culture and identified the most favorable locations in the United States for growing this crop. However, LeDuc was disappointed by the less than enthusiastic response of midwestern and northern agriculturists to this new potential source of sugar. He attributed this lackluster reception to the nature of root-crop cultivation—on hands and knees rather than on the comfortable seat of the then-common double-sulky cultivator.

5. William McMurtrie, *Report on the Culture of the Sugar Beets and the Manufacture of Sugar Therefrom in France and the United States* (Washington, D.C., 1880).

By late 1878 LeDuc's promotional efforts had turned from sugar beets to sorghum cane, a plant introduced into the United States before the Civil War. Sorghum became a focal point for USDA investigation for the next decade, since it offered midwesterners the opportunity to capitalize on the expanding home markets for sugar. Between 1878 and 1890 LeDuc and his successors at the USDA pinned their hopes on sorghum as the key source of sugar in their quest to create a home industry that would eventually result in American self-sufficiency in sugar production. LeDuc's decision to carry out research on sorghum sugar was not intended to discourage Louisiana tropical cane planters; rather, he hoped that sorghum could become a complementary crop that would extend the cane-growing and harvesting seasons, thus enabling the planters to realize a greater return on their capital investments in mills, clarifiers, and evaporators. The expensive sugarhouse machinery was usually employed only during the fall harvest between late October and early January for processing tropical sugarcane. Sorghum cane, with its early maturation, provided the planter with the additional opportunity to utilize his equipment between late August and early October. Yet the southern cane planters were doubtful of the success of sorghum, but remained confident in the future of their tropical canes. While sorghum containing up to 18 percent sugar had been cultivated, organic contaminants within the plant had made the crystallization process difficult and at times impossible. It seemed that while sorghum was a good source for molasses, it was a poor one for the production of crystalline sugar. Thus, the LSPA and a powerful congressional Louisiana delegation in Washington exerted sufficient political pressure to secure trials on tropical canes concurrent with the sorghum investigations.

LeDuc was convinced of the possibilities for sorghum as a national sugar crop during his visit to the 1878 Minnesota State Fair, when he obtained a sample of marketable sugar made by a Minnesota farmer using crude implements and apparatuses. After sending a sample to Washington for analysis, LeDuc decided to promote the crop, and he purchased a complete sorghum sugar apparatus for installation at the USDA laboratory in Washington. Several varieties of sorghum were grown on USDA grounds, and Peter Collier, McMurtrie's successor, directed the related analytical investigations. Collier's sorghum-sugar studies provided the basis for USDA sugar manufacturing trials conducted during the 1880s.

Peter Collier was born in 1835 in New York State. In 1861 he graduated from Yale, where he subsequently continued his chemical studies under the direction of Professor Samuel W. Johnson. After receiving a doctorate from Yale's Sheffield Scientific School in 1866, Collier was appointed to the Chair of Analytical Chemistry, Mineralogy and Metallurgy at the University of Vermont, where he also held a professorship in general chemistry and toxicology in the School of Medicine. In 1877 he left Vermont and became a chemist for the USDA, where he remained until 1883.[6] Between 1879 and 1882 Collier studied the commercial potential of sorghum cane as a sugar crop by examining the relationship between stages of growth and content of available sugar from different types of sorghum. The chief chemist concluded that some varieties contained sugar in quantities as high as that found in tropical canes cultivated in Louisiana.[7]

Working under newly appointed Commissioner of Agriculture George B. Loring, in 1882 Collier undertook several studies, including the determination of sugar content in different varieties of sorghum, the evaluation of the diffusion process for the extraction of sugar from sorghum cane, and the application of successive treatments of excess lime followed by the addition of carbon dioxide for clarifying extracted sorghum juices. Collier's diffusion studies revealed several difficulties in attempting to remove sugar from sorghum by osmosis.[8] To begin with, he recognized that diffusion resulted in a sub-

6. *Appleton's Cyclopedia of American Biography*, VII, 356; Peter Collier, *Sorghum, Its Culture and Uses: An Address Delivered by Dr. Peter Collier Before the Chamber of Commerce of the State of New York, March 5, 1885* (New York, 1885); Peter Collier, *Sorghum as a Source of Sugar: Including a Review of the Bulletins of the Department of Agriculture* (n.p., n.d.); Peter Collier, *Report of Analytical and other Work Done on Sorghum and Cornstalks* (Washington, D.C., 1881); Peter Collier, *Sorghum, Its Culture and Manufacture Economically Considered as a Source of Sugar, Syrup, and Fodder* (Cincinnati, 1884).

7. U.S. Department of Agriculture, *Report of the Commissioner of Agriculture, 1880*, 12.

8. U.S. Department of Agriculture, *Report of the Commissioner of Agriculture, 1881 and 1882*, 11; *Report of the Commissioner of Agriculture, 1883*, 7–8; *Report of the Commissioner of Agriculture, 1884*, 12, 16. On Loring, See Greathouse, *Historical Sketch of the U.S. Department of Agriculture*, 16; *Appleton's Cyclopedia of*

stantial dilution of extracted juices, compared to that incurred with milling extractions. This led to higher fuel costs during the evaporation step because of the increased percentage of water in the juice. Second, these diluted juices made defecation difficult, since a thinner blanket of scum was produced with the addition of lime. Third, in contrast to the dry bagasse by-product from milling, the exhausted diffusion chips were still moist after removal from the diffusion cell and were useless as fuel. Fourth, the thin juice was subject to a greater risk of inversion because of the increased time in the evaporation kettle. Finally, Collier found that sorghum juice contained large concentrations of organic materials, particularly gums, which were impossible to remove and resulted in a darker, lower quality product. Yet Collier was convinced that the successful application of the diffusion process was crucial to the economic feasibility of sorghum-sugar manufacture. His preliminary studies revealed that approximately one-third of all sugars in sorghum were left in the bagasse after milling. Collier had employed a series of interconnected barrels as his diffusion vessels, and he remained optimistic that the use of properly designed equipment would eliminate many of the shortcomings associated with his early trials.

In 1882 Commissioner Loring requested that the National Academy of Science (NAS) conduct a review of the USDA's sugar research program in general, and Collier's methods in particular. Since a committee of elite agricultural scientists advocated the use of chemical control in sorghum investigations and suggested the economic feasibility of sorghum cane as a sugar crop for Louisiana and other geographic areas, the NAS report had a far-reaching impact upon future USDA research.

The 1883 report, entitled *Investigation of the Scientific and Economic Relations of the Sorghum Sugar Industry*, may have been less than totally objective in evaluating Collier's work, since three mem-

American Biography, XV, 349–50; George B. Loring, *The Sorghum Sugar Industry: Address . . . Before the Mississippi Valley Cane Growers' Association, Saint Louis, Mo., December 14, 1882* (Washington, D.C., 1883). A short history of USDA sugar experiments to 1886 is contained in Norman J. Colman to Randall L. Gibson, May 26, 1886, in Record Group 97, Records of the Bureau of Agricultural and Industrial Chemistry, NA.

bers of the committee—William H. Brewer, Samuel W. Johnson, and Benjamin Silliman—were all from Yale University and may have been reluctant to harshly criticize one of the school's most successful alumni. Nevertheless, the NAS committee reviewed and approved USDA analytical methods and praised the laboratory's organization and division of labor—the chemists each specialized in certain analytical techniques but systematically cross-checked one another's results. In the opinion of the committee, future research should be centered upon fertilizer selection and proper dosages, methods of defecation, studies of new sorghum varieties, and manufacturing processes. In conclusion the committee remarked that "the Department of Agriculture, with its varied resources, scientific skill, mechanical appliances, and extended correspondence, coupled with the enormous circulation of its publications, can do this work as it cannot be done elsewhere." Collier viewed the NAS report as an endorsement not only of his research program at the USDA, but also his views concerning sorghum cane as the potential source of all the nation's sugar. Collier proclaimed "that the sorghum plant is destined, sooner or later, to furnish not only all the sugar needed in this country, but also a very considerable proportion of that required by foreign nations."[9]

Although the committee intended the report to be a full endorsement of Peter Collier's research program, it became a strong argument for continued sugar research at the USDA, even without Collier's leadership. Loring forced Collier to resign in 1883, apparently as a result of a long-standing personality conflict. In his place Loring appointed Indiana chemist Harvey W. Wiley. Never forgetting his dismissal, Collier used his political connections, and those of his brother-in-law, James B. Angell, president of the University of Michigan, to constantly campaign for his reinstatement at the USDA. As a result, Collier and Wiley emerged as bitter opponents. In contrast to both Collier and McMurtrie, Wiley displayed a remarkable sensitivity to regional interest groups like the Louisiana Sugar Planters' Association. His position in Washington, ever-threatened by the po-

9. National Academy of Sciences, *Investigations of the Scientific and Economic Relations of the Sorghum Sugar Industry, . . . November, 1882* (Washington, D.C., 1883), 25; Collier, *Sorghum, Its Culture and Manufacture,* 41.

litical machinations of his predecessor, would in time be strengthened by the support of the Louisiana sugar planters and their legislative representatives.

Like Collier, however, Wiley held to a vision of an America self-sufficient in sugar production. Throughout the 1880s Wiley maintained that a viable sugar industry was a primary national interest, and that this trade would stimulate business in other manufacturing areas. Wiley linked the prosperity of the sugar industry to the welfare of the American working man.

> I, for one, shall work to get that country [U.S.] to make his [the laborer's] life less burdensome, to give him his days' labor and not take it from him to build up all about him a varied industry. . . .
>
> If, in doing this, some coolie under a tropical sun should get less work and make less money, I will still my conscience by remembering that it is less cold there, the plumber and the coal dealer are less insistent. Sullenly lolling in the sun, he can look up and see cocoanuts [sic] and bananas; he will not starve or freeze.[10]

Harvey Wiley was born in 1844 and graduated from Hannover College in 1867. After earning a degree in medicine in 1871 at Indiana Medical College, Wiley enrolled at Harvard's Lawrence Scientific School, where he obtained a bachelor of science degree in 1873. He then returned to Indiana, where he became professor of chemistry at Purdue, as well as the Indiana state chemist. At Purdue, Wiley became interested in the sorghum-sugar industry and established himself as a scientific authority in this rapidly growing industry between 1879 and 1882.[11]

As chief chemist of the USDA in 1883, Wiley continued the investigations of sorghum-sugar manufacture initiated by Collier. However, under Wiley's direction the government experiments exhibited a marked departure from established USDA chemical and engineering methods used in the previous administration's programs. Unlike his predecessor, Wiley relied heavily upon the European beet-sugar industry and systematically assimilated apparatuses and techniques from this source into the USDA experiments conducted in Louisiana

10. Harvey W. Wiley, *Our Sugar Supply: Annual Address of President of the Chemical Society of Washington* (Washington, D.C., 1887), 33–34.

11. *Dictionary of Scientific Biography*, XIV, 357–58.

and the Midwest. Wiley acquired his knowledge through reading foreign treatises and international trade journals, meeting with representatives from European foundries, traveling to sugar factories abroad, and hiring individuals intimately familiar with beet-sugar technology.

Wiley's research into cane- and sorghum-sugar manufacture, which began in 1883, met with a series of failures and concluded in 1887 with only one noteworthy success in Louisiana. Wiley's ability to survive the calls for his resignation between 1886 and 1887 was attributable to the tight alliance he had painstakenly developed between the LSPA and USDA. On the whole, the LSPA viewed Wiley with favor, not only for assuming an ideological stance close to their own (like many planters, he advocated the protection of all home industries), but also because of Wiley's discrete use of appointment power within his own agency. He had also shown a willingness to include the association in his decision making. Wiley had skillfully built a network of personal alliances within the state of Louisiana, and LSPA members who opposed his experimental program found themselves first isolated within the membership and then crushed in firmly worded, authoritative rebuttals. Wiley's program to develop the American sugar industry was his first crusade, and one in which politics and science were inexorably entwined. Indeed, his experiences during the 1880s would prepare him for later battles that made his name a household word.

One of Wiley's first decisions as chief chemist in 1883 was to hire Guilford Lawson Spencer, a former pupil. Born in 1858, Spencer obtained a bachelor of science degree at Purdue in 1879, and then a master's degree from the University of Michigan in 1882.[12] Wiley needed Spencer, but not just to perform routine analysis; Spencer was one of the few chemists trained in American universities who had extensive experience in the European beet-sugar business. In 1882 and 1883 Spencer, on his own initiative, had worked in France, and in a letter to Wiley in the summer of 1883 he detailed his experience.

12. *American Men of Science* (1921 ed.), 646. Guilford Lawson Spencer, *A Handbook for Chemists of Beet Sugar Houses and Seed Culture Farms: Containing Selected Methods of Analysis, Sugar House Control, Reference Tables* (New York, 1897).

In Europe five weeks work in Mr. H. Pellet's Laboratory Paris—One campaigne [*sic*] in sugar manufactory at Francières (Oise) France, where I did *all* the chemical work for several weeks. Have done the work required in all positions from the lowest to superintendent. Have repeatedly performed the operation known in France as the "Cuite en Grain" or boiling to grain.

Understand the management and care of vacuum pans. As soon as the next season opens I will visit works in France and Belgium especially to see new methods in operation; also to see diffusion apparatus. I have already visited some of the most complete works in France.[13]

In May of 1883 Spencer began supplying Wiley with information on new apparatuses and processes. In addition, Spencer recommended to his former professor one of the most important new treatises on sugar manufacture—Paul Horsin-Dèon's *Traité Théorique et Pratique de la Fabrication du Sucre—Guide du Chemiste-Fabricant.*[14] Horsin-Dèon had been the personal secretary to Norbert Rillieux, and his treatise marked a departure from previous work in the field of chemical and sugar technology. It was one of the first studies that seriously attempted to apply the laws of physics to the efficient design of large-scale manufacturing equipment. Prior to 1883 Harvey Wiley was a bench-top analytical chemist; he was soon to be initiated into a new and sometimes strange world of chemical reactions on the large scale, and his scientific knowledge was only of limited use.

By midsummer Wiley had obtained an appointment for Spencer as assistant chemist at the USDA, but Spencer waited until the end of October to return to Washington. He remained in Europe to study new methods of filtration. The significance of Spencer's practical experience to the sugar experiment program during the 1880s cannot be underestimated. Spencer was able to combine a knowledge of chemistry with an understanding of engineering principles acquired from working in the sugarhouse. His confidence in his own engineering abilities was reflected in a July 22, 1883, letter to Wiley in which he asserted, "I am certain that I could give directions for

13. G. L. Spencer to H. W. Wiley, July 21, 1883, in Harvey W. Wiley Papers, Box 15, Library of Congress, hereinafter cited as LC.

14. Spencer to Wiley, May 17, 1883, in Wiley Papers, Box 15, LC; Paul Horsin-Dèon, *Traité Théorique et Pratique de la Fabrication du Sucre: Guide du Chemiste-Fabricant* (Paris, 1882).

having small diffusion apparatus made at a much smaller cost than you could obtain *experimental* apparatus in Europe."[15]

Another source of information for Wiley concerning beet-sugar technology in general, and diffusion in particular, was his correspondence with New Orleanian Rudolph Sieg. Like Collier, Wiley held that diffusion, rather than milling, was the key to establishing a profitable sugar industry in the United States. Sieg had been an important figure in the ill-fated experiments of the Julius Robert Diffusion Company in Louisiana during 1874 and 1875. In September of 1883 Wiley wrote Sieg expressing his surprise at the scale of experiments conducted in Louisiana during the 1870s. In reply, Sieg sent printed articles concerning the application of the diffusion process to the manufacture of cane sugar at Aska and answered several questions concerning Louisiana sugar manufacturing that were later reprinted in Wiley's 1883 report entitled *Diffusion: Its Application to Sugar-Cane, and Record of Experiments with Sorghum in 1883.*[16] Sieg's positive relationship with Wiley later proved to be crucial to the latter's continued popularity in Louisiana during periods of intense criticism of the USDA's sugar research effort. Also, Sieg, a knowledgeable and experienced engineer, routinely informed Wiley about technological developments abroad, translating German articles and describing engineering innovations in terms that Wiley could understand. Sieg was influential in LSPA circles, and particularly among Republican party members, including D. D. Colcock, secretary of the Louisiana Sugar Exchange, former Governor Warmoth, and a merchant group that included J. B. Levert and Louis Bush.[17]

Assisted by Spencer, Wiley began his diffusion studies in November, 1883, using a small cane cutter and a diffusion battery prototype manufactured by the Colwell Iron Company of New York. These experiments emphasized the use of systematic chemical analysis in the evaluation of the pilot-plant trials for studying the extraction of sugar

15. Spencer to Wiley, July 22, 1883, in Wiley Papers, Box 15, LC.

16. H. W. Wiley to Rudolph Sieg, September 25, 1883, in RG 97, NA; Harvey W. Wiley, *Diffusion. Its Application to Sugar-Cane, and Record of Experiments with Sorghum in 1883*, USDA Division of Chemistry Bulletin II (Washington, D.C., 1884), 20.

17. "Petition to Benjamin Harrison Requesting the Appointment of H. C. Warmoth as Collector of Customs," 1890 (MS in Warmoth Papers, Reel 7, SHC).

from sorghum. Diffusion and mill juice were compared by measuring the resulting solution's specific gravity, total solids content, total sugars, percent sucrose, ash, nitrogen albumen, and amides. An initial attempt was made to formulate a mass balance of the process by the analyses of waste waters, pulps, and semisyrups. Wiley also continued Collier's studies on the process of carbonatation, a beet-sugar clarification step that involved the treatment of juice with an excess of lime, and the subsequent removal of unreacted lime with carbon dioxide. One problem in the manufacture of sorghum sugar was the large percentage of sucrose remaining in the syrup. It was hoped that carbonatation would remove the impurities that were thought to hinder the crystallization process. Wiley was careful to point out that these initial tests were important to the cane-growing interests in Louisiana as well as to the sorghum interests. Wiley maintained that the chief purpose of investigations by the Bureau of Chemistry was to set up a rational, systematic procedure for the diffusion process and thereby convince planters of its inherent advantages over milling operations.[18]

By late 1883 the Division of Chemistry, under Wiley's leadership, was doing more for the Louisiana sugar industry than merely writing encouraging statements in USDA *Reports* and *Bulletins*. Spencer was sent to conduct chemical analyses at Henry Clay Warmoth's Magnolia plantation in Lawrence, Louisiana, during the grinding season of 1883–1884.[19] The decision to send Spencer to Warmoth's plantation was perhaps the result of the former governor's influence in Washington. Warmoth had close connections with the Chester A. Arthur administration. Yet, politics aside, Warmoth's plantation was the most outstanding sugar factory on the banks of the Mississippi; for its sugarhouse contained the latest designs of process equipment, and its manufacturing efficiency was a frequent topic of conversation within the local planter community. Perhaps as a consequence of a tumultuous and controversial political career, Warmoth redirected his talents and energy into the business enterprise that became second to none in Louisiana by the mid-1880s. The Warmoth estate drew the admiration of Mark Twain who, in *Life on the Mississippi*, incor-

18. Wiley, *Diffusion. Its Application to Sugar Cane*, 20.
19. Spencer to Wiley, January 14, 1884, in Wiley Papers, Box 15, LC.

rectly but comically described the complex technology found at Magnolia.

> The great sugar-house was a wilderness of tubs and tanks and vats and filters, pumps, pipes and machinery. The process of making sugar is exceedingly interesting. First you heave the cane into the centrifugals and grind out the juice; then run it through the evaporating-pan to extract the fiber; then through the bone-filter to remove the alcohol; then through the clarifying-tanks to discharge the molasses; then through the granulating pipe to condense it; then through the vacuum pan to extract the vacuum. It is now ready for market.[20]

Warmoth's interest in sugar manufacturing began during the 1870s when he entered into a partnership with Effingham Lawrence of Plaquemines Parish. By the early 1880s Warmoth had become extremely interested in employing chemical analysis to monitor changes in sugar processing. And he introduced new designs of mechanical apparatuses to enhance manufacturing efficiency. Warmoth also hired chemical and engineering experts from New York. One of these was A. Lavandyra, whom Warmoth employed as a consulting chemist in 1881 when studies were conducted on the chemical composition of both extracted process juices and sugarcanes in the field.[21] Warmoth was particularly interested in the effects of various fertilizers on the quality of the cane; to that end he had applied different fertilizers and manures in separate areas on the plantation and subsequently evaluated their merits through the use of comparative analytical data. These 1881 studies showed that a combination of cottonseed meal and peavines applied to the cane seemed to be more effective than either phosphate, guano, cottonseed, or peavines alone. For engineering knowledge, Warmoth relied on inventor and engineer Samuel Fiske of New York City.[22] In 1884 Warmoth installed Fiske's cane shredder for the purpose of increasing extraction efficiency. The experiments that followed proved rewarding; Warmoth's extraction efficiency with an imperfect mill averaged 74.6 percent during 1884 and 1885.

20. Samuel L. Clemens, *Life on the Mississippi* (New York, 1917), 384.

21. "Juice Analysis," November 26 [1881?], "Proposed Manure for Cane" [1881?], "Record of Analysis, December 6, 1881; all in Warmoth Papers, Reel 4, SHC.

22. A. Lavandyra to H. C. Warmoth, May 20, 1884, Reel 4, and H. C. Warmoth to Samuel Fiske, January 18, 1884, Reel 9, all in Warmoth Papers, SHC.

However, Warmoth's friendship proved to be both a blessing and a liability to Wiley. Perhaps the most capable businessman-planter in Louisiana and certainly a powerful advocate of new technology, Warmoth had many enemies. Although a friend of both Chester A. Arthur and Benjamin Harrison, he was resented by those Republicans who were once close to Ulysses S. Grant and detested by Democrats like Randall Gibson. Despite the fact that Warmoth was on excellent terms with Duncan Kenner, Wiley had to be careful in granting favors to Warmoth.[23] With each concession given to the former governor, Wiley in turn had to equally placate Warmoth's opponents in Louisiana.

In order to continue his experiments in 1884, Wiley was dependent upon appropriations from Congress. The appropriation bill for 1884–1885, as passed in the House, contained no provision for USDA studies in the manufacture of sugar; however, Senator Randall Gibson, an LSPA member, along with Senator Preston B. Plumb of Kansas, added an amendment to the bill in the Senate, calling for $50,000 for the general expenses of the laboratory and sugar experiments. This proved to be a windfall for Wiley, since laboratory expenses were only $10,000, and he was thus left with $40,000 for additional manufacturing trials.

Because of the late passage of the bill, Wiley and Commissioner Loring were faced with the difficult task of contracting for experimental machinery and devising meaningful experiments in time for the fall harvest. In March of 1884 Wiley contacted the Pusey and Jones Company of Wilmington, Delaware, about building a suitable diffusion apparatus.[24] When the LSPA learned of the appropriation, its leaders began to establish ties with Wiley, who they perceived to be in control of future investigations. On March 14, 1884, Henry Studniczka of the LSPA informed Wiley that "a committee was appointed to invite you through Gen. Loring to come down here and lecture to the planters of this state. . . . A lecture on Diffusion will be very acceptable as the question is being strongly agitated in this state."[25]

23. Henry Clay Warmoth, *War, Politics and Reconstruction* (New York, 1930), 81.
24. D. D. Cone to Wiley, March 19, 1884, in Wiley Papers, Box 15, LC.
25. H. Studniczka to Wiley, March 14, 1884, in Wiley Papers, Box 15, LC.

Wiley clearly saw the advantage of cultivating the LSPA's support. In early April he wrote Commissioner Loring:

(1) . . . I think it highly important that the results of the experiments of the Department with diffusion should be known in Louisiana where such great benefit would accrue by the introduction of this process.

(2) I do not know what prospect there is of securing from Congress an appropriation for further experiments in this line. I should like very much during the next campaign to have the department establish a station in Louisiana and to make a thorough study of the chemistry of cane and the processes of manufacture.

(3) The active aid of the planters of Louisiana would doubtless be of great help in securing such national protection and aid for our struggling sugar industries as their importance demands.[26]

During the spring of 1884, Wiley established contacts with several influential members of the LSPA. Although unable to attend the opening exercises of the Louisiana Sugar Exchange, he engaged in correspondence with that institution's secretary, D. D. Colcock. J. L. Brent wrote Wiley in July, 1884, assuring the chemist that the association would assist the USDA in any way possible. Wiley replied by emphasizing the application of his current sorghum experiments to problems in cane-sugar manufacture, and stated that the USDA diffusion apparatus would be on display at the upcoming World's Industrial and Cotton Centennial Exposition in New Orleans. Wiley closed his letter to Brent by stating, "I am in most hearty sympathy with the sugar industry in Louisiana, and while I occupy my present position will do all in my power to aid and encourage it."[27]

Eager to conclude his tenure as commissioner with a successful demonstration of the diffusion process, Loring contracted with the Pusey and Jones Company on July 14, 1884, for the necessary experimental apparatus. From the beginning, the venture was destined for failure. The Wilmington foundry had no clear understanding of the intended use of the machinery and had access only to outdated designs for diffusion cells. Successful diffusion required that the cane

26. Wiley to George Loring, April 4, 1884, in RG 97, NA.

27. D. D. Colcock to Wiley, June 19, 1884, in Wiley Papers, Box 16, LC; Joseph L. Brent to Wiley, July 12, 1884, in Wiley Papers, Box 15, LC; Wiley to J. L. Brent, July 16, 1884, in RG 97, NA.

be cut into very thin sections. However, in 1884 there existed no adequate mechanical models to serve as a basis for the construction of a workable cutter. By September the diffusion vessels were complete and approved by the USDA, but the inadequate cane cutters were refused. Clearly, it was impossible to ready the apparatus for November trials in Louisiana, and the newly purchased apparatus was reluctantly put in storage.

The agency's inability to properly guide Pusey and Jones in the design of the experimental apparatus showed USDA leaders that they lacked essential knowledge concerning the mechanical aspects of sugar making. Steps were taken to remedy this weakness by sending qualified individuals to study European designs. In mid-July of 1884 the federal government decided to sponsor travel abroad for Warmoth. Fred T. Frelinghuysen, secretary of state in the Chester A. Arthur administration, provided the appropriate diplomatic introductions.[28] Wiley's assistant chemist, Spencer, accompanied Warmoth in these European travels. During the trip Spencer informed Wiley: "[I] Have spent all my time working the French Chemists for all they are worth—Called on Mr. Pellet a couple of times. . . . We would have left Paris sooner had we known that the sugar houses had commenced work—From now on we shall have our hands full—Governor Warmoth wishes to sail for America about the middle of next month."[29]

During the fall of 1884 the USDA mounted several small-scale studies in Louisiana. After returning from Europe, Spencer conducted experiments at Warmoth's Magnolia plantation, while chemists Charles Crampton and J. I. Donahue were stationed at the USDA's exhibit at the World's Industrial and Cotton Centennial Exposition in New Orleans and John Dymond's Belair plantation, respectively.

Spencer began his work at Warmoth's plantation on November 11, 1884. He evaluated a cane shredder and its effect on extraction efficiencies, and conducted studies of various methods of filtration. Cane was passed through a shredder, and the resulting pulp was conveyed first through a three-roller mill and then immediately between the rollers of a twin-cylinder device. Spencer performed analyses of

28. Warmoth, *War, Politics and Reconstruction,* 261.
29. Spencer to Wiley, September 11, 1884, in Wiley Papers, Box 16, LC.

the cane, the extracted juices, scums, syrups, sugars, and bagasse. Samples were also sent to Washington for subsequent albuminoid determinations. Evaluations of the efficiency of the process were performed by assaying the raw and clarified juices. The loss of sugar caused by the disposal of kettle skimmings and clarification scums in the open kettle method was a major focus of Spencer's investigations. It was known that these process impurities contained large quantities of sugar, and Spencer showed that by passing a slurry of these waste products through a filter press more than 80 percent of the sucrose could be recovered and returned to the process stream. Spencer's European experience proved invaluable in these tests, since the apparatus employed was a Riedel design manufactured in Germany. His work at Magnolia, however, was not confined to the sugarhouse; he also examined the effect of cottonseed and superphosphate fertilizers upon the growth and quality of the cane. A second USDA chemist, Charles Crampton, was located at the Cotton Centennial Exposition in New Orleans, where he performed analyses for planters upon request, and also provided instruction in the use of the polariscope.

Donahue, less experienced than Spencer or Crampton, conducted routine analyses on mill and clarified juices and syrups at Dymond's plantation. In addition, he analyzed several varieties of cane, producing data that revealed a much lower sucrose content in tropical cane than Wiley had expected. Donahue summed up his work at Belair by stating: "From the experience I have had this season I find that a chemist ought to be a sugar maker as well in order to secure a good place on a plantation, then the chemist can be head man. I find that the sugar makers are not very willing to instruct one in their business."[30] USDA chemists employed in Louisiana were quickly discovering that chemical knowledge alone was insufficient to control sugarhouse processes. The manager of a modern installation required a unique combination of chemical and mechanical expertise. However, the effectiveness of the scientist's efforts was dependent upon his authority. Chemists like Spencer and Donahue were enmeshed in a struggle for power within the sugarhouse, their adversaries being the rule-of-thumb sugar boiler and mechanic. Until

30. J. I. Donahue, to Wiley, January 10, 1885, in Wiley Papers, Box 18, LC.

the chemists could display practical abilities at least equal to that of the traditional artisans, their ideas would count for little in the real world. The application of science to the improvement of sugar processing could be accomplished only when the chemist held a supervisory position.

With the data from these three preliminary, independent investigations, Wiley concluded that the small mill and open kettles must be abandoned in Louisiana if the local industry was to remain competitive. Further, he felt that the diffusion process had to be given a fair trial in the near future. He closed his summary of the recent experiments by asserting:

> I had expected to find the mean percentage of sucrose in the juices of cane at least fourteen and was not a little surprised to find it greatly less.
>
> One of the great problems to which the sugar-cane grower should seriously address himself is to secure the production of a cane richer in sugar. Careful and systematic selection of seed, and a constant practice of a most favorable system of fertilizing and cultivation will surely result in such an improvement. . . .
>
> The best way to accomplish this result would be the establishment by the State of an experiment station where a principal object of the work would be the improvement of the quality of the cane.[31]

Wiley's suggestions were taken seriously by the leaders of the Louisiana Sugar Planters' Association. In the spring of 1885 preliminary steps were taken to establish a sugar experiment station from funds raised by private subscription.

While the three chemists were busily engaged in analytical laboratory activities, Wiley firmly established the alliance between the United States Department of Agriculture and the LSPA during his address to the association's December, 1884, monthly meeting. The chief chemist's visit to New Orleans coincided with the opening of the World's Industrial and Cotton Centennial Exposition; a USDA exhibit at the exposition was a reflection of the federal government's role in building a new South. The fairgrounds were located on land that was once the Boré plantation and the site of the fa-

31. Harvey W. Wiley, *The Sugar Industry of the United States,* USDA Division of Chemistry Bulletin V (Washington, D.C., 1885), 69.

mous eighteenth-century "experiment" immortalized by Charles Gayarré. The main building stood near the center of the exposition site, and it contained mechanical equipment including steam engines, cotton compresses, and a small diffusion battery operated by USDA chemists.[32]

Rather than give the planters in the audience the impression that a visiting scientist possessed a know-it-all attitude, Wiley began his speech on a humble note. He stated that he was in Louisiana to gather information and not to tell intelligent men who had invested the better part of their lives in the manufacture of sugar how to run their business. Turning to statistics on the European beet-sugar industry, Wiley described the rise of international competition, a theme similar to that employed by John Dymond at other, less publicized, LSPA meetings. The chief chemist maintained that Americans had no need to purchase imported sugar from either Europe or the Caribbean. The sugar industry should be solely contained within the borders of the United States, and it should be protected by a tariff or supported by a bounty. "Not only the sugar growers, therefore, but all agriculturists and all engaged in every kind of industry should unite to oppose bringing the agriculture and manufacture of this country into competition with the cheap labor of other lands. What this country can produce, and that is almost everything, should be produced here. I should like to see the duty on sugar increased, or some system of bounty inaugurated by which, in a few years, we would make all the sugar we consume."[33]

Wiley had struck the right nerve as far as the planters were concerned, since his views were remarkably similar to those propounded months earlier by the LSPA leadership at the Louisiana Convention for Protected Industries. Wiley had said what the planters wanted to hear; it was not surprising for prominent LSPA member T. S. Wilkinson to remark that the speech was "the most interesting address ever delivered before this Association."[34] Thus the relationship be-

32. D. Clive Hardy, *The World's Industrial and Cotton Centennial Exposition* (New Orleans, 1978).

33. Harvey W. Wiley, "The Sugar Problem," in Louisiana Sugar Planters' Association, *Regular Monthly Meeting, December 11, 1884* (n.p., n.d.), 12.

34. *Ibid.*, 15.

tween the United States Department of Agriculture and the Louisiana Sugar Planters' Association was cemented, and Wiley had gained the acceptance of Louisiana's most influential sugar planters. The chief chemist would soon need his newfound friends to support him in scientific and political battles centered in Washington.

7

The Great Experiment

Harvey Wiley's speech at the Louisiana Sugar Exchange in December, 1884, marked the beginning of a fruitful and lasting relationship between the United States Department of Agriculture and the Louisiana Sugar Planters' Association. For the remainder of the decade the sugar bowl experienced a rebirth. While northern capital and the energy of planters played an important role in this revitalization, the presence of USDA chemists and the encouragement to install advanced-design equipment and laboratory controls contributed significantly to this transformation. The LSPA leaders felt that their best chance to secure the assistance of the federal government lay with Harvey Wiley, and out of their own interests they ended up supporting this controversial chief chemist.

Even individuals outside the planters' organization recognized Wiley's favored position. By the end of 1884, Commissioner Loring's term was almost over, and one of the strong contenders for the position was Norman Jay Colman of Missouri. Colman, a lawyer turned publisher, had strongly advocated high protective tariffs for the American sugar industry in his journal, *Rural-World*. Throughout the 1880s he propounded a policy of self-sufficiency in sugar production. Further, he felt that Louisiana would continue to play a prominent role in supplying the United States with sugar. In a letter to Wiley, he inquired: "Can't you get [the] Louisiana Sugar Planters Association to take some action in the matter [his own appointment]? I would probably suit them better than most Democrats. My *speech* shows that I am for protecting our industry."[1] Apparently Wiley followed up on Colman's inquiry, since in the spring of 1885 J. Y. Gilmore, an LSPA member, wrote to Wiley that "our association could not endorse *any* one for Commissioner, because its Charter prevents its taking *any* political action. However, Mr. Colman is the choice of most members."[2]

A similar view of the LSPA's unwillingness to support a candidate for commissioner of agriculture was expressed by John Dymond, who wrote Wiley in March, 1885. "We had our meeting of the Planters' Asscn, and I consulted some of the leading gentlemen there & there seemed an indisposition to take any political action, as the main purpose of our organization is to advance our cause agriculturally & to consider the immediate political status of sugar. Even the latter is considered of doubtful profit & the . . . political action we developed here last spring, was under the auspices of a convention of sugar planters & entirely apart from the Planters Asscn."[3] The LSPA leaders were hesitant to politically endorse a candidate for government service because they realized that the organization was a loose-knit coalition of Democratic and Republican interests. A political move, even if in the interest of Louisiana planters, could upset a delicate equilibrium. Politics was an accepted means for the association's efforts only in the case of tariff legislation, where both groups worked together for the common good of tariff protection.

Despite a policy of noncommitment, a number of agricultural groups from other parts of the country supported Colman. As a result of this grass-roots campaign, Colman was successful in receiving the appointment of commissioner of agriculture during Grover Cleveland's first administration. Perhaps because of Wiley's strong support of his new boss, the chief chemist had an inordinate amount of influence within the USDA between 1885 and 1888.

In the meantime, Loring, in his last months of office, decided that sorghum experiments, rather than investigations into the manufac-

1 Norman J. Colman to H. W. Wiley, December 24, 1884, in Wiley Papers, Box 16, LC.

2. J. Y. Gilmore to Wiley, March 18, 1885, in Wiley Papers, Box 16, LC.

3. John Dymond to Wiley, March 25, 1885, in Wiley Papers, Box 16, LC.

ture of cane sugar, should be continued during the coming fall of 1885. The commissioner instructed Wiley to travel to New York City, where he was to arrange a contract with Benjamin Urner of the Parkinson Sugar Company of Ottawa, Kansas, for the installation of an experimental diffusion battery at that company's factory. Unhappy with the commissioner's decision, and without the commissioner's knowledge, Wiley contacted Warmoth in January of 1885 about the proposed USDA program. The chemist was sure that success depended upon the integration of the diffusion process with an efficient evaporation apparatus, and he suggested to Warmoth "that it would be well for the Louisiana planters to take steps to secure the location of the battery in that state, where such production facilities as I have spoken of can be readily secured. . . . I do not wish to appear as a meddler in the business affairs of the department—but—I have so much interest in the development of the sugar industry that I do not like to see such a costly and important experiment fail for lack of proper conditions." [4]

As a result of Wiley's letter, LSPA representatives contacted Loring and requested the installation of a similar experimental apparatus in Louisiana. Loring agreed, and the Pusey and Jones Company was awarded a contract on February 27, 1885, to furnish the USDA with a cane cutter and twelve diffusion cells arranged in a circle, for delivery not later than June 15, 1885. Unlike the first Pusey and Jones apparatus contracted by the USDA in 1884, the new equipment was based on Guilford Spencer's designs, and it was hoped that "the battery . . . should have all the advantages of the latest European apparatus." [5]

By mid-March the LSPA's major topic of discussion was the intended location of the USDA experiments in Louisiana. While Wiley and others at the USDA may have favored Warmoth's plantation, it was clear that the LSPA leadership did not. It was highly probable that while the Republican members of the association favored Warmoth's Magnolia plantation, the Democrats, including Dymond and Senator Gibson, favored Kenner's Hermitage plantation, which was said to be more accessible by boat. In the end, the LSPA execu-

4. Wiley to H. C. Warmoth, January 28, 1885, in Warmoth Papers, Reel 4, SHC.
5. Wiley to William G. Gibbons, March 16, 1885, in RG 97, NA.

tive committee, authorized by the USDA to make the final decision as to the experiment's location, selected Kenner's plantation. Kenner, satisfied with this arrangement, instructed Senator Gibson to recommend to Colman, the new commissioner of agriculture, that Wiley be appointed in charge of the Louisiana experiments.[6] At last, Wiley had sole responsibility for the federal government's sugar experiments—something he had been unable to achieve under the Loring administration.

Upon assuming office in April, 1885, Colman's first crisis was the discovery that the USDA's operating funds for the fiscal year ending in June, 1885, were nearly exhausted. Through departmental mismanagement, an inordinate amount of the agency's appropriation ($30,000) had been spent by Loring for the procurement and distribution of sugar-beet seeds. Contractors for experimental machinery ordered by the department could not be paid, and installations of the equipment in Kansas and Louisiana were delayed. However, Senators Gibson and Plumb pushed through the Senate another generous appropriation for fiscal year 1885–1886, and by early summer a start was made in erecting the experimental apparatuses. Colman appointed M. A. Scovell of the Illinois Industrial College to oversee these installations in Wiley's absence. Scovell had prior experience in operating Illinois sorghum factories. By July he was busy supervising construction on Kenner's plantation.

Because of its presence in Louisiana, the USDA was flooded with requests for associated appointments. For example, Henry Studniczka, who had asked for an appointment as chemist in 1884, renewed his efforts through the offices of Congressman Andrew Gay, an LSPA member. Wiley was apparently unimpressed with Studniczka's abilities, but he diplomatically wrote Gay: "I am aware that Mr. Studniczka has had a large experience in the conduction of diffusion experiments. . . . If after work begins I find there is a lack of experience and skilled labor I shall be very glad to take advantage of the experience and knowledge which Mr. Studniczka possesses." However, Wiley did perceive that Rudolph Sieg could be of value to him in the proposed trials. In July, 1885, Wiley considered sending Sieg to

6. Duncan Kenner to Wiley, March 25, 1885, in Wiley Papers, Box 18, LC.

Brazil to witness attempts to apply diffusion to the extraction of sugar from cane.[7]

Wiley also wanted to ensure that Sieg's friend, Warmoth, would continue to receive USDA assistance, even though the LSPA committee had rejected Magnolia plantation as the experimental location for the evaluation of the diffusion process. He wrote to Warmoth in August. "I shall try and arrange affairs so that the station for the chemical examination of sugar juices may be continued at your plantation during the coming season. I am much interested in the outcome of your work for the next campaign for if you can secure an average expression of 80 percent of juice without too great an expenditure of power it will certainly prove of the greatest benefit to the sugar interest."[8]

With consummate skill, Wiley concurrently worked to ensure that Warmoth's neighbor and rival, John Dymond, would not feel snubbed by the Division of Chemistry. When Dymond finally submitted a request for the services of a USDA chemist, Wiley replied that since the privately sponsored Louisiana Sugar Experiment Station was only beginning its operations, Donahue would again be assigned to Belair to assist Dymond. Wiley optimistically claimed that "much can be done toward helping the sugar interests by such work as you ask for."[9]

Wiley's plans for studies of the process of diffusion in Kansas and in Louisiana met with unexpected disasters during the early fall of 1885. Pusey and Jones were late in shipping the necessary machinery, both to Ottawa and to Hermitage. Finally, in late September the last carload of equipment arrived in Ottawa, Kansas. Wiley and his USDA colleagues rushed to prepare the experiments, while Scovell remained in Louisiana awaiting instructions for the final preparations there. Again, to his disappointment, Wiley found that the Pusey and Jones equipment was totally inadequate for the intended task. Instead of having followed Spencer's designs, the Wilmington foundry had relied on their own imperfect knowledge of

7. Wiley to Andrew Gay, July 15, 1885, Wiley to Rudolph Sieg, July 29, 1885, both in RG 97, NA.

8. Wiley to Warmoth, August 3, 1885, in RG 97, NA.

9. Wiley to Dymond, October 30, 1885, in RG 97, NA.

sugar technology. Among the defects was a smaller-than-proposed discharge hole on the diffusion cells, a design shortcoming that required workmen to tediously dig out the exhausted chips rather than quickly discharge them from the vessel's bottom, as planned. In addition, the USDA contract with the foundry had called for each cell to have a capacity of two tons; in fact, each diffuser held only fourteen hundred pounds. Because of these defects, Wiley realized that all hope for a successful trial in Louisiana had vanished. He wrote Scovell on October 9, 1885, to postpone the experiments and to store the machinery until the following season.[10]

Although the apparatus at Ottawa was far from satisfactory, Wiley carried out some limited experiments, focusing primarily on the process of carbonatation. Wiley concluded that diffusion and carbonatation ought to be employed together, and that it was feasible to extract 95 percent of the available sugar from the sorghum plant as a marketable product.[11]

As soon as Wiley returned to Washington, Colman instructed him to visit Europe "for the purpose of inspecting and purchasing such forms of machinery as may appear most useful for the purpose named [for an investigation of the diffusion process], and to gather such information in respect of the sugar industry as may be calculated to secure the greatest success in the work which the Department has undertaken." The LSPA membership wanted to make sure that its interests would be served during Wiley's European trials. Therefore, on November 23, 1885, Kenner requested that D. D. Colcock, secretary of the Sugar Exchange, use his influence with Colman to see that Wiley visited Almeria, Spain, where the French Fives-Lilles Company was erecting a diffusion plant for cane-sugar manufacture. In addition, the LSPA requested that Wiley publish a report as soon as possible detailing the apparatuses and processes employed in Almeria. Fully aware of the less-than-satisfactory results obtained at Ottawa in 1885, the cancellation of the proposed diffusion trials on Kenner's Hermitage plantation, and the continued loyal support of Senator Gibson in Congress, Colman was eager to comply with

10. Wiley to M. A. Scovell, October 9, 1885, in RG 97, NA.

11. Harvey W. Wiley, *Experiments with Diffusion and Carbonation at Ottawa, Kansas. Campaign of 1885*, USDA Division of Chemistry Bulletin VI (Washington, D.C., 1885).

Colcock's wishes, primarily to retain LSPA support. The commissioner wrote Colcock that "I am so deeply interested in these investigations, however, and so anxious to do everything in my power to encourage the cause which your association represents that I have today directed Prof. Wiley to comply with your suggestions, and have sent him your letter for his information and guidance."[12]

Wiley examined large sugar refineries, first in Britain then in France. In Paris he met with D. Lucassen, an influential planter from Java who was instrumental in the Javan diffusion tests on sugarcane in 1885. From his conversation with Lucassen, Wiley became convinced that the diffusion cells designed and fabricated by the French Fives-Lilles Company were superior to those of its competitors, and in compliance with the wishes of the LSPA he arranged to witness the operation of this equipment at Torre Del Mar and Almeria, Spain.

Wiley's close observations of the machinery employed at Almeria equipped him with the knowledge necessary to avoid many potential engineering problems with the future USDA trials. While in Spain, Wiley realized that book knowledge alone was not sufficient to solve the manufacturing problems that had thwarted his past efforts at diffusion and would surely confront him in the immediate future. He wrote to Colman that "the most beautiful theory requires a skilled hand to prove effectual in practice and that the diffusion of cane is a business *sui generis* and must be learned by experience."[13]

Although Spencer and Warmoth had made previous trips to Europe to study sugar manufacturing, and Sieg possessed over a decade of experience related to the diffusion process, it was not until after Wiley's 1885 trip to Europe that the USDA systematically applied the fundamentals of beet-sugar technology to the production of cane sugar. During 1886 and 1887 virtually every piece of apparatus employed in USDA trials was based upon European designs. Diffusion cells and calorizators (heaters) were manufactured according to drawings supplied by the Fives-Lilles Company. Cane cutters were purchased by the USDA from either the Sangerhäuser Maschinenfabrik or the Südenberger Maschinenfabrik of Magdeburg, Germany. Lime

12. Colman to Wiley, November 19, 1885, in Wiley Papers, Box 15, LC; Colman to D. D. Colcock, November 25, 1885, in Record Group 16, Records of the Commissioner of Agriculture, NA.

13. Wiley to Colman, December 16, 1885, WP, Box 16.

kilns used in the carbonatation experiments and the sulfur kilns for clarification studies were fabricated according to instructions from Sangerhäuser. In addition, the tanks, pans, and pumps necessary for carbonatation were direct copies of a system manufactured by the Hallesche Maschinenfabrik of Halle, Germany. To complete the carbonatation step, filter presses separated the resulting juices and scums, and even this equipment was of German origin.[14]

During Wiley's travels in Europe, Spencer was gaining practical experience in sugar technology at Magnolia plantation in Louisiana. Between late 1885 and early 1886, Spencer continued his earlier studies in which he attempted to increase the sucrose content in canes by the application of chemical and natural fertilizers. He also documented the relative quantities of sugar extracted from canes of different maturities. Employing both physical and chemical analysis, Spencer calculated the efficiency of Warmoth's mill. He also examined the chemical reaction of inversion, or the degradation of sucrose to glucose, which often took place during the clarification step. Using a Hallesche Maschinenfabrik filter press, Spencer analyzed the effluent from the filters and the resulting by-product press cake. Spencer's evaluation of the furnace temperatures of a Fiske bagasse burner was the most innovative experiment conducted at Magnolia between 1885 and 1886. Although similar experiments were being conducted by British scientists, the idea of measuring the temperature at various locations within a furnace was unique for an American manufacturer of the period.

Both the USDA and Magnolia plantation were held in high regard in Louisiana prior to 1886. The publication of the report detailing the successful USDA experiments at Magnolia, as well as Warmoth's efficient and progressive management, resulted in international acclaim for both the department and Warmoth. Commenting on the high yield of sugar extracted from Warmoth's milling operations, one contributor to the widely circulated British journal *Sugar Cane* commented that the proper management of a mill was perhaps preferable to the use of a diffusion battery. "It is interesting to note how

14. Harvey W. Wiley, *Methods and Machinery for the Application of Diffusion to the Extraction of Sugar from Sugar Cane and Sorghum, and for the Use of Lime, and Carbonic and Sulphorous Acids in Purifying the Diffusion Juices*, USDA Division of Chemistry Bulletin VIII (Washington, D.C., 1886), 56.

close Magnolia has come to this limit [the maximum extraction of sugar from a roller mill]; with the plant they have at present I consider this work quite a revolution in the manufacture of sugar. Advocates and makers of diffusion machinery, pause! Confess you have been to [sic] hasty,—rather go to Ex-Governor Warmouth [sic], at Magnolia, and say, we have sinned in the sight of the public, and before thee, make us thy 'boss weighers' for the coming campaign."[15]

While Warmoth was drawing such praise in early 1886, Wiley was assailed with criticism. Throughout 1886 and 1887, Wiley's detractors gloated at his reversals, and nearly caused his dismissal or resignation. However, the alliance between the LSPA and the USDA ultimately thwarted those who felt that Wiley was either incompetent, impractical, dishonest, threatening, or personally objectionable.

The most vocal and politically powerful of Wiley's opponents was Congressman T. Floyd King from Vidalia, Louisiana. King was born in Georgia, educated at the University of Virginia, and served in the Confederate army during the Civil War. He relocated to Vidalia, a town across the Mississippi River from Natchez, in 1872 and started a law practice, later becoming a respected planter. He served in Congress between 1879 and 1886.[16] Perhaps because it was an election year and because USDA presence was in south Louisiana rather than north Louisiana, King was looking for controversy and spoiling for a fight. In a letter to a New Orleans newspaper in February of 1886, he initiated a two-year battle with Wiley. His comments were directed against Wiley's integrity and his scientific and technical abilities. King claimed that the beet-sugar industry was based upon principles dating from the Napoleonic period and that only practical men, not scientists, were capable of succeeding in the sugar business. King saw no need for trips to Europe and considered them to be a waste of the taxpayers' dollars. His ideas reflected the strong anti-European sentiment that was expressed by many Americans in the 1880s. In March, 1886, King suggested to Kenner that Wiley be relieved of future responsibility in the USDA sugar experiments.

I clearly see that unless Professor Wiley, . . . is removed from his present position, the hope of getting further appropriation from Congress for

15. J. Owen Alexander, "Diffusion vs. The Work Done at Magnolia Plantation, Louisiana," *Sugar Cane*, XVIII (1886), 540–41.

16. *Biographical Directory of the American Congress, 1774–1971*, 1236.

this worthy purpose will be greatly imperiled. . . . Do you know of any one who would fill this position? Some time ago Commissioner of Agriculture [Colman] stated to me that if I would name a suitable person he would have him appointed. . . . Mr. Gay tells me he is unable to find the man needed. I therefore write to you to ask that you take steps at once to furnish me the name and address of such a person. . . .

This does not require the qualities of a chemist so much as those of a practical constructive sugar engineer and chemist. It is simply arranging the most simple machinery and putting it into operation.[17]

Kenner, along with others within the LSPA, had worked hard to develop close ties with the USDA and accepted none of King's arguments. He defended not only the disbursement of USDA appropriations, but also Wiley's personal qualifications.

He impressed me very favorably. He certainly appeared to understand the scientific view of the application of the diffusion process to the tropical sugar-cane. Of course he is not a mechanic, but he even understood the mechanical part of the process better than could be expected. The Government has paid, I have no doubt, several thousand dollars to have him instructed in the matter of diffusion, and it seems to me it would be bad policy to displace him, and to put another in his place to serve. . . .

I know several chemists who have the necessary scientific knowledge to qualify them to discharge the duties of the position, but who have had no practical experience of the mechanical difficulties to overcome. . . . I therefore can recommend no one to take his place, but advise that Professor Wiley be allowed to finish the experiment which he has commenced.[18]

Kenner's opinion of Wiley was in part influenced by conversations with Sieg.[19] However, the LSPA foundrymen, who were outside the association's circle of leadership, apparently agreed with King. Charles G. Johnsen, a practical engineer and representative of the Reading Iron Works, joined with King in calling for Wiley's ouster. It is likely that these so-called millmen felt that their livelihood would be jeopardized by the replacement of the roller mill with foreign-made diffusion cells. Johnsen wrote William H. Hatch on March 16, 1886.

17. J. Floyd King, The Government Experiments and the Sugar Trade (Washington, D.C., 1887), 11–12.

18. Ibid., 12 (a reprinted letter, Duncan Kenner to J. Floyd King, March 15, 1886).

19. Sieg to Wiley, March 15, 1886, in RG 97, NA.

In a review of the work done by the Agricultural Bureau I do not find a single discovery of any value in the sugar industry; on the contrary, much harm has been done by the misleading statements of the bureau.

The Government has spent $218,000 attempting to educate Professor Wiley as an engineer and sugar maker. It is a soft thing for Wiley.

The solution of the future of the sugar industry does not rest, in the remotest degree, with the Bureau of Agriculture.[20]

Even though Johnsen and other millmen openly opposed the USDA's presence in Louisiana, they had little, if any, impact upon LSPA policy. Diplomacy and negotiations on higher levels would ultimately be decisive in determining the future of the Division of Chemistry's experimental program. In addition to Kenner's support, Wiley also received the endorsement of LSPA vice-president John Dymond. Dymond wrote King in April, 1886, stating that any change in personnel at this crucial juncture would further delay a program from which planters were anxiously awaiting results.[21]

While Rudolph Sieg had influenced Kenner, Dymond's support had been carefully cultivated by John Harper, an old friend of Wiley's from their days at Purdue University. In 1886 Harper was the chief engineer of the Shell Beach Railroad near New Orleans. He used this position to "say some things, incidentally, with more effect than might otherwise come from another source." King's position with the sugar planters was further eroded because of his negative stance on the Morrison bill, a measure which would have maintained the protective duties on sugar. Wiley stood for protection, while King did not, and the LSPA would rather support a Republican chemist from Indiana advocating a protectionist philosophy similar to that of the membership than a north Louisiana Democrat with conflicting interests. In early July, Wiley sent several documents on diffusion to his friend Harper, who distributed them to local planters. Before the summer was over, King's threatened disruption of the program appeared to subside. In mid-July Harper wrote Wiley stating: "It is in the best hands now, and I think it wiser for me to step aside and not champion the thing, but let it take its course. King is unpopular here

---

20. King, *The Government Experiments and the Sugar Trade*, 24 (a reprinted letter, Charles G. Johnsen to William H. Hatch, March 16, 1885).

21. *Ibid.*, 16 (a reprinted letter, Dymond to King, April 25, 1886).

now on account of his action on the Morrison bill. What I want, is to have the planters endorse your actions." And by early August, King was convinced that at this stage of the battle Wiley's influence over the sugar planters and the LSPA remained strong. Nonetheless, King warned Dymond that he would have to continue his fight to sustain the chief chemist.[22]

Wiley decided to concentrate USDA experimental efforts at Fort Scott, Kansas, during the fall of 1886. However, in addition to the sorghum trials, Louisiana cane would be shipped north at the end of October for processing with the same experimental apparatus. To witness these experiments, the LSPA sponsored three representatives, two of whom were advocates of Wiley's program—Rudolph Sieg and E. C. Barthelmy, a New Orleans sugar maker. The third member, W. P. Kirchoff, a consulting sugar chemist and refiner, was openly critical of Wiley's methods. Anxious to demonstrate the effectiveness of his department and its ability to wisely allocate the appropriated funds for applied research, Colman regarded the Fort Scott tests as crucial. He wrote Wiley on September 25, 1886.

> I have a notion to blow you up. It may be you have been blowed up already. I think some great calamity must have befallen you or I should have heard from you. I have been waiting day after day, and week after week, I will not say patiently, but impatiently to hear what was going on at your sugar factory. Not a word can I hear, however, from anybody whether you are at work or not in making sugar. I don't know as yet whether a single ton of cane has been worked. I want to know just how you are getting along and what the prospect is; how the machinery is working and what are the drawbacks. I have very great anxiety over the result of your campaign.[23]

The Fort Scott trials employed the latest designs of European apparatuses and relied on the strict use of chemical control in the sugarhouse. Assistant chemists in the USDA performed routine analysis on the diffusion juices, filtered carbonated juices, sulfurated liquors, waste waters, exhausted chips, the semisyrups, the massecuites,

22. John Harper to Wiley, June 28, 1886, Harper to Wiley, July 26, 1886, both in Wiley Papers, Box 19, LC, see also William McKinley, Jr., to John Dymond, May 3, 1886, in Dymond Family Papers, Folder 35, HNOC. King to Dymond, August 3, 1886, in Joseph L. Brent Papers (copy), Box 6, Louisiana State Museum.

23. Colman to Wiley, September 25, 1886, in Wiley Papers, Box 19, LC.

product sugar, and molasses. In addition, analytical chemical techniques monitored the acidity of extracted juices, the limestone used in the carbonatation kilns, and the by-product press cakes. Congressman Gay's sugarcane was similarly extracted by the diffusion method and subsequently treated by the carbonatation process. The chemical analyses revealed that the diffusion cell temperature of 70°C, commonly used for sorghum processing, was too low for tropical cane. However, at 90°C extraction efficiency was improved.

As a result of the Fort Scott tests, Wiley was convinced that diffusion and carbonatation, when conducted together, resulted in a 30 percent increase in extraction efficiency over that obtained by traditional milling practices in the manufacture of sugar from sorghum. Wiley postulated that carbonatation would increase the purity of the product sugar by reducing the glucose concentration, eliminate losses through skimming, and promote crystallization in the coolers. In his report to the commissioner, Wiley asserted that the next logical step in the USDA's experimental sugar program was the establishment of a Louisiana experiment station to employ the Fort Scott extraction and clarification apparatuses, as well as a superior double- or triple-effect evaporator.[24]

Wiley's report failed to convey the numerous difficulties encountered by the USDA personnel at Fort Scott. To begin with, Wiley was surrounded by his critics, who had come to Fort Scott for the sole purpose of witnessing what they hoped to be an embarrassing manufacturing failure and who questioned the chemist's every move. In the midst of these difficulties, Wiley wrote: "You can't imagine what a h—l of a time I have had here. The devil himself in the person of P. C. [Peter Collier] has been here and had had a crowd of imps with him. Some of the persons whom I have most benefited joined the enemy supposing I would fail with Louisiana cane. My glorious victory here has completely discomfited the enemy and all they can do now is to misrepresent. But the truth will triumph in the end."[25]

Wiley's "glorious victory" was a matter of opinion, and it appears

24. H. W. Wiley, *Record of Experiments at Fort Scott, Kansas, in the Manufacture of Sugar from Sorghum and Sugar-Canes, in 1886*, USDA Division of Chemistry Bulletin XIV (Washington, D.C., 1887), 52, 56.

25. Wiley to Orville D. La Dow, November 17, 1886, in Wiley Papers, Box 19, LC.

that only the chemist himself viewed the Fort Scott experiments with such enthusiasm. From December of 1886 to June of 1887, his opponents accused Wiley of a number of serious improprieties, causing many to question his integrity and his competence. He remained in office primarily because of steadfast support from Commissioner Colman and influential LSPA members.

One of the first reports describing the Fort Scott experiments appeared in the St. Louis *Globe Democrat* on October 29, 1886. The special correspondent asserted, "The answer to this [failure at Fort Scott] is a long story involving the usual inefficiency of the agents of the Government, and demonstrating the advantage which practical businessmen have over the mere theorist."[26] During the experiments Wiley was apparently engaged in a controversy concerning his expertise and authority with USDA employee Magnus Swenson, a former professor at the University of Wisconsin and an experienced sugar maker. Press reports claimed that any success at Fort Scott was due to Swenson's efforts and not Wiley's. Rather than continue with Swenson's suggested approach, Wiley had insisted on employing hot water diffusion on sorghum. This method failed to produce a crystallizable sugar. However, during Wiley's brief absence Swenson successfully applied cold water diffusion and produced a high yield of sugar from sorghum. The press viewed this conflict as the Washington theorist versus the midwest practical man and used the incident as the basis for a sensational story. The Fort Scott *Daily Monitor* of November 5, 1886, remarked: "The so-called experiments have been conducted by a government chemist a great part of the time without even the presence of a practical sugar maker on the premises. These very excellent gentlemen know all about oxygen, hydrogen, carbonates, chlorides, ihynol, eucalyptol, and are probably familiar with diastase, but can not tell with certainty when a pan of sugar is ready to strike without making a chemical analysis."[27] The local newspaper concluded with a statement that reiterated the chief contention of the St. Louis *Globe Democrat*—namely that private enterprise, rather than the federal government, should provide the leadership

26. Reprinted in King, *The Government Experiments and the Sugar Trade*, 19.
27. Fort Scott (Kans.) *Daily Monitor*, November 5, 1886, in Louisiana Sugar Industry Scrapbook, Louisiana Collection, Howard-Tilton Library, Tulane University.

of future research in the sugar industry. Many of the press reports concerning the experiments at Fort Scott were circulated widely throughout the United States and were received with particular attention by many in Washington. These reports would later form the basis of a second round of intense criticism of Wiley's integrity and expertise.

It appears that Wiley, at this juncture in his career, was incapable of managing large-scale manufacturing trials. Although he thoroughly understood the analytical chemistry related to sugar manufacturing, he was unable to successfully combine chemical and mechanical knowledge. Wiley's inexperience in the sugarhouse resulted in an overall poor showing at Fort Scott. Even two of his closest supporters, Rudolph Sieg and Sangerhäuser Maschinenfabrik representative Ernst Schulze, disagreed with Wiley's yield calculations. However, they decided that it was in the best interests of the sugar industry not to publish their findings.[28] While Sieg and Schulze expeditiously chose to disregard these discrepancies, Wiley's most ardent opponent in the LSPA, W. P. Kirchoff, was not so compliant. However, at the November and December, 1886, monthly meetings, other influential LSPA members thwarted Kirchoff's designs. D. D. Colcock, secretary of the Louisiana Sugar Exchange and close friend of Rudolph Sieg, successfully proposed to the LSPA that Sieg's report of the Fort Scott experiments be published by the association, while Kirchoff's summary be tabled. Spencer, who was assigned to Magnolia plantation after leaving Fort Scott, informed Wiley of Colcock's efforts to suppress Kirchoff's report.

> I called on Mr. Colcock this P.M. He expressed himself very strongly in regard to Kirchoff. It is sufficient to say that he hasn't a very high opinion of him. At the last meeting of the association Colcock played a neat trick on Mr. K. The report of Mr. Sieg as published in the papers was under consideration. Colcock moved that this report be considered an official report to the Association. . . . The Association adopted the report and ordered it published in the next number of the bulletin. Mr. Colcock says that at the next meeting he will move that Kirchoff's report be tabled.[29]

28. Sieg to Wiley, November 26, 1886, in Wiley Papers, Box 19, LC.
29. Spencer to Wiley, November 18, 1886, in Wiley Papers, Box 19, LC.

J. B. Wilkinson, Jr., son of a prominent Plaquemines Parish planter and editor of the New Orleans *Item*, proved to be another important member of the pro-Wiley circle in Louisiana. In a letter written to Wiley in early December, Wilkinson suggested that Wiley "better not antagonize the mill men . . . by not claiming more than the figures show." However, even though the accuracy of Wiley's data was questionable, Wilkinson reassured the chemist. "You probably are aware of Mr. Kirchoff's animus against you and this is to say that I have interested myself to . . . [arrange] for this report, which will be read tonight, a suitable reception. You may rely on having friends to look after your interests at this end of the line."[30]

Even though he was supported by his Louisiana friends, Wiley was discouraged at the controversy surrounding his work. He wrote to a friend on November 27, 1886: "You will know the bitter and malicious fight, which Collier, Parkinson, Swenson & Co. are making against me. . . . It is entirely probable that I will have nothing whatever to do with the Louisiana experiments, and may even be asked to sever my connection with the department."[31]

Wiley recognized that Commissioner Colman's support was the key factor that allowed him to maintain his USDA appointment under these trying circumstances. Colman had received numerous letters condemning Wiley's approach at Fort Scott, and the latter defended himself vigorously. Wiley had been accused of dishonesty by a representative from Pusey and Jones. Wiley responded to Colman.

> Replying further to the two atrocious letters of Mr. M. Day, I can only say that they are the expressions of a man neither honorable nor truthful.
>
> I have never, in this country or Europe, received one cent directly or indirectly, from any source as commissions on apparatus bought or for any other cause whatever. The only bribe that I was ever offered was before I was appointed to my present position. . . .
>
> No baseless slander can long injuriously affect an honest and upright man.[32]

30. J. B. Wilkinson to Wiley, December 9, 1886, in Wiley Papers, Box 17, LC.

31. Wiley to Blackwell, November 27, 1886, in Wiley Papers, Letter Press Book 4, LC.

32. Wiley to Colman, December 3, 1886, in Wiley Papers, Letter Press Book 4, LC.

While it is doubtful that Wiley was deliberately falsifying experimental data, it is apparent that his mechanical competency proved to be difficult for Colman to defend. From eyewitness accounts at Fort Scott, Wiley was apparently incapable of uniting chemical and engineering knowledge for application to practical manufacturing problems. Commissioner Colman and W. L. Parkinson exchanged a series of letters that detailed his shortcoming. Parkinson asserted:

And yet it will not be amiss to refer to the statement in your [Colman's] letter that "both Mr. Potts and Mr. Hughes say that to make a sugar engineer, a sugar boiler and a chemist all out of one man when they were distinct occupations, was more than should have been expected." This statement exposes an error into which you have fallen. . . . If any man undertook and assumed all these "distinct occupations" what occasion was there for it? And if he failed in the assumed task, who is to blame? And if there was failure from this source shall we be able to learn from it? Before the season began I wrote you suggesting the importance of having a sugar boiler and a skilled man to run the centrifugal machines, and proposed that the Dept. join us in securing them. . . . The plain truth is that Dr. Wiley assumed all those positions and many more. Let us not deceive ourselves. The failure, so long as there was failure, was not for want of competent help, nor was it for want of a good crop.[33]

Colman made the decision to keep Wiley as chief chemist, but in order to win support for future USDA endeavors, he allowed the LSPA additional influence in decision making. During December, USDA officials were uncertain that LSPA support for their program would be forthcoming. Spencer wrote to Wiley in late December, 1886.

The Planters are becoming a little restless about the work to be done here; the talk is that all but $20,000 has been spent in fact Mr. Dymond told me that he had received information to that effect from Mr. Parkinson. I told Mr. Dymond that I did not believe that our expenditures at Fort Scott exceeded $45,000—There is also a fear that you are not hurrying forward the machinery as rapidly as possible. I hear a great deal for and against you and your management of the work. . . .

Quite a number of planters have expressed themselves as opposed to an experiment on a small scale. They ask me to use my influence.[34]

33. Parkinson to Colman, December 6, 1886 (copy), in RG 97, NA.
34. Spencer to Wiley, December 18, 1886, in Wiley Papers, Box 19, LC.

The first step in strengthening the alliance between the USDA and the LSPA took place in late December, when Commissioner Colman approved Congressman Gay's recommendation that E. C. Barthelmy be appointed superintendent of the upcoming diffusion experiments in Louisiana. Barthelmy, one of the LSPA observers at Fort Scott, was a practical sugar maker who had become Spencer's close friend. Since the LSPA insisted that the USDA conduct a large-scale diffusion trial, Spencer suggested that Warmoth's plantation was the only feasible site for the experiment. On January 1, 1887, Spencer wrote Wiley.

> The La. planters are only too ready to believe evil reports of you. You know they are not a class progressive, and rather gloat over the failures of new processes. You cannot afford more trouble such as has resulted from your Kansas work. Even if untrue these newspaper reports will surely damage you unless you can save yourself in Louisiana. Current report says you are going to erect a hundred and fifty ton battery on a plantation up the river. I would have favored such a place a few weeks ago, but now I know that it would be a great mistake.—Mr. Dymond, Mr. Thompson and a number of others I am certain would tell you the same thing. Some even go so far as to say, "better not make the experiment unless you can make it on a larger scale." Warmoth is ready and willing to back you in this work, but I fear that even he would condemn an experiment on a small scale.[35]

LSPA demands influenced not only Wiley's programs of research, but also Spencer's experiments at Magnolia during the 1886–1887 season. Although Spencer continued studies that he first began in 1884 (analysis of process fluids and solids and evaluation of the effect of specific fertilizers), he also assessed a filtration process at the request of D. D. Colcock. Ernst Schulze of the Sangerhäuser Company worked with Spencer in the appraisal of the Kleeman filtration process. After liming, juice was brought to a rapid boil, slurried with charcoal for ten to fifteen minutes, then passed through a Hallesche Maschinenfabrik filter press, where the color-causing scums were removed. Spencer found that the product sugar was of higher quality,

35. Wiley to Spencer, December 22, 1886, in Letter Press Book 4, Spencer to Wiley, November 18, 1886, in Box 19, Spencer to Wiley, January 1, 1887, in Box 17, all in Wiley Papers, LC.

less labor was required, and process sugar losses were kept at a minimum.

In January, 1887, the A. W. Colwell Company of New York received a contract to build the diffusion battery for the upcoming tests in Louisiana. The design was based on drawings furnished by the German Sangerhäuser Company. Meanwhile, Louisiana planters' interests began to shift away from perfecting the diffusion process to improving mill designs. It appears that only a small number of Louisiana planters, including several members of the LSPA leadership, were genuinely interested in introducing the diffusion process into Louisiana. The majority of planters were conservative in their views concerning new technology. Indeed, many gloated over the failures of the diffusion apparatus in Kansas. Many seemed unaware of the scientific and technical advances that were restructuring the international sugar market and threatening the very existence of sugar manufacturing in Louisiana. Proponents of milling techniques touted Spencer's recent success at Magnolia, where 78 percent extraction was obtained using the latest roller mill design. Public opinion may have been influenced by the efforts of J. B. Wilkinson, Jr., the editor of the *Item*, who attempted to assuage the anger of the Louisiana millmen against Wiley's diffusion program. In an article entitled "Let Moderation and Fairness Prevail," Wilkinson stated, "It is possible that the mode of extraction which is perfect for beet sugar may not be best for cane."[36]

Moderation, however, was not a trait of J. Floyd King. In March, 1887, he intensified his year-long battle against Wiley with a malicious personal attack during appropriation debates in the House of Representatives. King's speech, along with numerous letters and newspaper reports, was published by the Government Printing Office and distributed using King's postal frank. D. D. Cone, a representative of the Pusey and Jones Company, along with Peter Collier, ensured that this pamphlet enjoyed a wide circulation among the scientific and technical community. King's broadside was designed not only to denounce Wiley as an incompetent chemist and an incapable manager, but also to praise Collier's early sorghum studies. King

---

36. New Orleans *Item*, February 25, 1887, in Louisiana Sugar Industry Scrapbook.

reprinted an 1885 letter in which Senator Preston Plumb had expressed his negative opinion of Wiley's mechanical talents.

> I am quite strongly impressed with the belief that unless you [Pusey and Jones Company] send some practical person to Ottawa, Kansas, to superintend the setting up and operation of the machinery that you manufactured for the Government, the result will be failure. . . .
>
> The Government has a handful of persons employed, but, . . . [they] lack the proper qualifications. There is no interest in the experiment on the part of anyone in Government employ, and unless there is some change in the situation there will be simply the record of another dismal failure.

King carefully assembled numerous newspaper clippings, including articles from the St. Louis *Globe Democrat* and the New Orleans *Times-Democrat*, to reinforce Plumb's contentions. One report stated: "All agree that the experiments have been conducted extravagantly and unsatisfactorily: that Professor Wiley, while perhaps an able chemist, has not the practical knowledge of sugar-making to conduct the experiments, and that if results of any value are to be secured, future appropriations to test this important question must be under different auspices."[37]

However, King's printed attack on Wiley was not thoroughly convincing. In his attempt to include the letters of several prominent LSPA members, even King's clever editing could not mask the clearcut support given to Wiley by Dymond, Kenner, Gay, and Gibson. The best that King could do in using the statements of these leaders to promote his cause was to use selected quotations out of context.[38]

Although King employed the opinions of others in his printed attack, he constructed his own thesis for explaining the USDA's failures under Wiley's leadership. King postulated that Wiley had entered into a relationship with foreign machinery builders whose ultimate goal was to stagnate the American sugar industry. King contended that the European machine building trade had forged a strong alliance with a compliant Wiley. To support his argument, King cited a letter from the director of Fives-Lilles Company to Wiley, which suggested that he postpone the 1886 Louisiana experiments. Accord-

---

37. King, *The Government Experiments and the Sugar Trade*, 10, 21.
38. *Ibid.*, 24.

ing to King, within weeks of receiving this letter, Wiley decided to follow the Frenchman's advice. King's ideas concerning this conspiracy were expressed on the House floor.

> The influence that the English, German and French sugar interests are secretly exerting upon our affairs has not received the attention that it should. Indeed, it has scarcely been recognized. . . .
>
> When I read of the unsuccessful efforts made by agents of English sugar refiners in Parliament to obtain legislation adverse to the sugar interests of the United States, I at once perceived that an effort would be made to obtain secretly and by indirection in Washington what they had failed to obtain openly in London. . . .
>
> That he [Wiley] is, in my judgment, in the employ of French and German beet-root sugar manufacturers and sugar-machine builders, and London refiners, whose sole object is to stagnate all efforts of the Government to aid in developing our cane-sugar industry, by the introduction of the diffusion process which seems to me to be the only hope of our sugar planters. . . .
>
> That he [Wiley] is working in European sugar interests and against our own [is shown by] his every official and unofficial act since 1882.[39]

King was not opposed to the process of diffusion, but rather to Wiley's leadership in directing the government experiments and his use of foreign machinery. Intelligent observers may have considered King's accusations preposterous; and even if true, they came much too late to have any effect. By March of 1887 the USDA and LSPA regarded the upcoming fall experiments as a joint venture; an LSPA representative was to direct the trials at an experiment station chosen by the association. In addition, the LSPA was to serve in an advisory capacity. It was planned that at regular intervals a select group of planters would inspect the installation of the USDA apparatuses and would suggest future courses of action to the USDA scientists.

In early 1887 the USDA had planned to conduct the trials at the plantation of Judge Emile Rost, but this may have been merely a concession to placate Kenner, Rost's close friend. Months later an LSPA committee consisting of Kenner, Dymond, McCall, Wilkinson, L. C. Keever, W. B. Schmidt, J. C. Morris, and W. C. Stubbs concluded that because Warmoth's equipment and canes were the best in Louisiana,

39. *Ibid.*, 28.

Magnolia plantation "would afford the severest competitive test of any in the State." Wiley was extremely pleased with this decision. For over a year Sieg had been recommending Magnolia and Warmoth, claiming that the former governor was "not only the *best*, but *in fact* the *only* man in Louisiana *capable* of grasping the whole thing." Thus, Wiley was assured that the diffusion tests would be conducted on a plantation where interference from his enemies would be minimized and his chances for success would be maximized. As a reciprocal concession to the LSPA, Wiley promised D. D. Colcock, secretary of the Louisiana Sugar Exchange, that Rudolph Sieg would either be appointed as a USDA representative to observe diffusion trials at Demerara, or he would be considered for "any employment which we will be able to give him."[40] Thus, Wiley took every opportunity to insure that the LSPA had a vested interest in the outcome of the 1887 experiments at Magnolia plantation.

In mid-April of 1887 Colman and Warmoth signed the contract that established each party's obligations in the upcoming investigations. The USDA furnished Warmoth with the diffusion battery, carbonatation and sulfuration equipment, filters, sheds, and pumps with the understanding that they all remained the property of the USDA after the trial's conclusion. In addition, the USDA was responsible for the employment of all skilled labor and chemists.

For his part, Warmoth was obliged to provide all necessary space for the experiments, as well as sugarcane, steam, a steam engine to drive the experimental cane cutters, and unskilled labor. In the sugarhouse he was responsible for evaporation and crystallization of the product sugar from the juices obtained with the experimental apparatus. Further, Colman provided Warmoth with a guarantee that if he sustained any financial loss because of USDA-caused delay, he would be fairly compensated. In the event that Warmoth's claim was disputed, Kenner would appoint a three-man LSPA arbitration committee.[41]

As the consequence of these LSPA/USDA agreements, Wiley was

40. U.S. Department of Agriculture, Division of Chemistry, *Record of Experiments Conducted by the Commissioner of Agriculture in the Manufacture of Sugar from Sorghum and Sugar Canes . . . 1887–1888* (Washington, D.C., 1888), 79; Sieg to Wiley, May 5, 1886, Wiley to Colcock, May 19, 1887, both in RG 97, NA.

41. Colman to Warmoth, April 12, 1887, in Warmoth Papers, Reel 4, SHC.

certainly better situated to confront his relentless critics. After the widespread circulation of King's pamphlet, Wiley's opponents renewed their attacks. W. P. Kirchoff, who had claimed in December of 1886 that Wiley's Fort Scott results were "Among the Impossibilities," again questioned the analytical methods used at Fort Scott, the figures published in the Division of Chemistry's bulletin of the amount of cane ground, and the actual yield of sugar obtained by diffusion in the Kansas trials. Kirchoff's doubts were reiterated by Peter Collier, who in early May wrote the New Orleans *Times-Democrat* claiming that "Professor Wiley has in several instances shown inability to agree with himself even when discussing his own data, and his conclusions drawn from identical data have been flatly contradictory."[42] Collier cited examples of conflicting data from several of Wiley's USDA bulletins.

During the late spring of 1887, new antagonists emerged to criticize Wiley. Seth Kenny, the Minnesota farmer whose methods had convinced Commissioner LeDuc to study sorghum cultivation in 1878, asserted: "It is true that these experiments of Professor Wiley were absolutely worthless in establishing anything which was not already known to everybody else possessed of even a very limited knowledge of the subject under investigation. It is doubtful whether a chemist or a sugar maker could be found who would not a year ago have condemned . . . the 'process' [carbonatation] which Professor Wiley pursued during the entire season at Fort Scott."[43]

Kirchoff's attack, which previously had been aimed only at local audiences, was broadened to include international sugar interests with the publication of his Fort Scott report in *Sugar Cane* on June 1, 1887. In all probability, Kirchoff's arguments were far from being total distortions of the truth. Wiley had conducted some analyses improperly, and even his partisan supporters privately questioned the validity of his calculated yields. It must be remembered that the other two LSPA observers at Fort Scott (Sieg and Barthelmy) had close ties with the USDA. Further, Kirchoff's isolation within the LSPA was probably not so much a result of the singularity of his

42. New Orleans *Times-Democrat*, April 25, 1887, May 5, 1887, both in Louisiana Sugar Industry Scrapbook.

43. *Farmer*, May 19, 1887 (Poughkeepsie, N.Y.?), in Louisiana Sugar Industry Scrapbook.

ideas as it was a consequence of internal politics. Wiley's rebuttal of Kirchoff's claims, published in *Sugar Cane*, was primarily a character attack based on Wiley's perception of his and Kirchoff's relative stature within the American scientific community.

> In the locality where Mr. Kirchoff and myself are known, it would not be necessary to me to reply to such an accusation. . . .
>
> Mr. Kirchoff was sent last year, by private subscription, as a delegate from the Louisiana Planters' Association, to study the process of diffusion at Fort Scott.
>
> Instead of attending to his duties he busied himself with collecting materials for a violent and malignant abuse. . . .
>
> So disgusted were the members of the Association with his conduct, that, when his report was made to the Association, it was referred to a committee as the quickest method of burial. On the other hand, the reports of Messrs. Barthelmy and Sieg, his *confreres* at Fort Scott, were published in full. . . . Mr. Kirchoff's claim to scientific ability rests on grounds wholly unknown to the chemists of this country, among whom he is neither known nor recognized. Aside from this, however, no honorable man, however much he might be swayed by personal enmity, would consent to indulge in the species of misrepresentation which characterizes the whole of his screed [tirade].[44]

Wiley's only reference to Kirchoff's evidence—an estimation of sugar concentration in waste water—did not conclusively disprove the latter's claims.

Wiley's newfound assertiveness was not confined merely to defending himself against the criticisms of an unknown chemist; he also wrote a harsh letter to Senator Plumb. Reacting to the appearance of Plumb's letters in King's pamphlet, Wiley asserted that "this accusation [incompetency] coming from you does me a most serious injustice."[45] Perhaps the chemist's aggressiveness can best be understood in light of his renewed support from both Commissioner Colman and the placated LSPA membership.

Despite the numerous critics and portents of failure, the USDA forged ahead with its plans. Until Wiley arrived at Magnolia in October, Spencer, Barthelmy, and Warmoth made the day-to-day deci-

---

44. H. W. Wiley, "Diffusion at Fort Scott, Kansas," *Sugar Cane*, XIX (1887), 490–92.
45. Wiley to Senator Plumb, June 29, 1887, in Wiley Papers, Letter Press Book 4, LC.

sions. Although Barthelmy was the official superintendent of the experiment station, he deferred to Spencer in decisions concerning the placement of machinery, the repair of inadequate apparatuses, and the hiring and firing of personnel. By late July preparatory work at Magnolia was nearly finished. Foundations for the battery had been laid; the pit for the cane cutter was nearly finished; the carbonatation pans were in place and all connections made, except for those to the Kroog filter presses. Spencer had completed the fabrication of a lime kiln and had overhauled the process pumps. The sulfur apparatus was also ready, along with the engine for driving the cutters. All that remained was the completion of the building, but this could not be done until the diffusion cells were installed.

News of the total failure of the trials at Demerara reached Wiley in late July.[46] Since the failure was attributed to the inadequate Sangerhäuser cane cutters, Wiley immediately wrote to the foundry for assistance in modifying the USDA apparatus. The company replied that a German machinist would be sent to Louisiana to perform any necessary alteration. However, as a precaution, Wiley contracted with the Südenberger Company to manufacture an alternate cutter based on the reportedly successful design employed in Java.[47]

After Barthelmy suffered a hernia in September, Spencer was solely responsible for the erection of the buildings and the initial inspection and testing of the recently received diffusion battery. In a progress report to Wiley, dated September 29, 1887, Spencer displayed his knowledge of the apparatus.

> We are at present making a test of the battery. The hydraulic joints have been tested to 33 lbs. water pressure. In making the test one broke, several leaked and one flange was broken in tightening up to stop a leak. . . . This part of the apparatus now easily stands 30 lbs. pressure. We have been prevented today from trying the battery on account of a very heavy rainstorm. . . . It will take several days to stop leaky joints. . . .
>
> Let me offer a suggestion in filter pressing based on my experience. A pump working direct does not give a sufficiently steady stream of juice. I suggest that you authorize me to provide the arrangement described on page 15 [in USDA blueprints]. . . .

46. Ernst Schulze to Wiley, July 28, 1887, in RG 97, NA.
47. Wiley to E. C. Barthelmy, September 8, 1887, in RG 97, NA.

The damper regulator opens and closes the steam inlet to pump. The weight on lever regulates pressure on presses & this apparatus is perfectly automatic.[48]

Upon further testing, Spencer concluded that the diffusion cells were fabricated with inferior materials and poor workmanship. However, with the assistance of local foundrymen, the battery was repaired, pressure tested, and readied for the upcoming tests.

Wiley arrived in Louisiana in mid-October, and though not officially in charge of the investigations, he appears to have dominated all future work there. The Magnolia experiments, which had been the focus of controversy for nearly a year, almost ended prematurely on the night of October 19 when a hurricane struck Plaquemines Parish. The storm tore down one end of the USDA building, but fortunately the machinery escaped damage. After several days of frantic repairs, the diffusion trials were set to begin.

Initial newspaper reports of the work at Magnolia were encouraging. J. B. Wilkinson, Jr., reported in the *Item* that the USDA scientists had attained a 98 percent extraction efficiency. However, all journalistic accounts glossed over the formidable problems encountered by Wiley and his USDA colleagues.[49] Unable to handle the sometimes crooked cane stalks, the cane cutter choked at each irregularity. Unless enough cane could be fed through the cutters, the resulting low production capacity would overshadow the high extraction efficiencies obtained. Thus, Warmoth instructed his field hands to carefully trim the cane, making sure that all irregular pieces of cane were hand cut at the problem-causing bends. Wiley was certain that this bottleneck spelled doom for the experiments, and he wrote to Colman.

I have done everything possible to get our apparatus in working order— but—I have lost all hope of being able to do so.

I shall try for a day or two longer and then stop all fruitless attempts and put the machinery in order to be condemned and sold. . . .

The causes which have led to failure here are numerous . . . I have spent

48. Spencer to Wiley, September 29, 1887, in RG 97, NA.
49. New Orleans *Item*, November 19, 1887, and November 20, 1887, and New Orleans *Picayune*, November 22, 1887, all in Louisiana Sugar Industry Scrapbook.

two months of the most terrible anxiety that any one has ever experienced.—I only wonder that I retain my reason.

The best we have ever been able to do has been 30 tons per day.

A day later Wiley wrote:

I do not see much of encouragement in the present state of affairs. . . . I have at last got the cutter in condition to cut at the rate of 150 tons in 24 hours. The battery too is doing better but is still far from satisfactory. Yesterday in 5 hours we worked 30 tons of cane and got an extraction of 97%. Yet you must remember that . . . [we have] to clean the battery out every night. . . .

The reason we are not cutting more per day is on account of the inability to get the cane to the cutters.

Labor is very scarce here. . . .

We are now preparing to make our test run. Gov. Warmoth will shut down his house and give us a fair chance. We will invite Mr. Dymond & Mr. McCall to be present so that they may make an official report to the Sugar Association. Gov. Warmoth will prepare for us 150 tons of cane—perfectly clean and cut at the elbow—(all the cane is crooked and for this reason hangs in the hoppers). This will be placed near the cutter so that eight men can keep it going. I believe we will get through with this in 24 hours.

To-day and to-morrow I will finish all the experiments with carbonatation and lignite.[50]

Between December 3 and December 5 the mechanical problems associated with the cutters and diffusion cells were finally corrected by foundry representatives. The apparatus was able to process 80 tons of cane per day with an extraction efficiency of 226 pounds of sugar per ton of cane. The large-scale production of cane sugar by the diffusion process was suddenly a reality. However, because of unforeseen expenditures for equipment repair, USDA funds were totally depleted. This financial crisis threatened to end the trials just as they were beginning to show promise. However, Warmoth offered to continue USDA work with his own funds, provided that Commissioner Colman request Congress to fully reimburse him. The LSPA mem-

50. Wiley to Colman, December 2, 1887, in Wiley Papers, Box 23, LC; Wiley to Colman, December 5, 1887, in RG 97, NA.

bership was also persuaded to support continued USDA experimentation. Meeting in mid-December, the LSPA passed a resolution to petition Congress, through Senator Gibson and Congressman Gay, for $10,000 to cover the additional work at Magnolia.

With LSPA encouragement, Wiley continued to study the diffusion process at Magnolia. In a series of five tests, the capacity of the apparatus was gradually increased to process over 400 tons of cane per day, while obtaining an extraction efficiency of over 98 percent. These results proved conclusively that diffusion held a competitive edge over traditional milling techniques by more than 30 pounds of sugar per ton of cane. In a subsequent USDA bulletin describing the work at Magnolia, Colman asserted that "the diffusion process for the manufacture of sugar has advanced beyond the experimental stage by the labors of this Department, . . . and it is now offered to the sugar growers of the country with the confident assurance that it is the best, most simple, and most economical method."[51]

The United States Department of Agriculture and the Diffusion Process in Louisiana, 1888–1900

Wiley's victory at Magnolia could have marked a beginning by the United States Department of Agriculture to further improve sugar manufacturing processes. Instead, the department adopted a policy that assisted the Louisiana sugar industry in an indirect rather than a direct manner, leaving the task of perfecting the mechanical processes to private industry and to the Louisiana Sugar Experiment Station. However, Colman and Wiley did not totally abandon their once-important planter allies. It was agreed that the large planters would employ the assistant chemists, on leave without pay from the USDA, to assist them during harvest season in the installation and operation of diffusion equipment. In 1890 chemists were employed at A. R. Shattuck's Des Lignes plantation, J. N. Pharr's Glenwild plantation, and L. S. Clarke's Lagonda. During this period, Wiley's nephew, assistant chemist Hubert Edson, worked for Wilbray Thompson at his Bayou Teche sugar factory and frequently wrote to Washington for scientific and technical assistance.[52]

51. U.S. Department of Agriculture, *Record of Experiments . . . in the Manufacture of Sugar from Sorghum and Sugar Canes*, 4.

52. See W. J. Thompson, "Brown Coal and Wood Char in the Filtration of Cane

The USDA also funded experimental sugar investigations. Between 1888 and 1900 the USDA closely cooperated with the scientific staff at the Louisiana Sugar Experiment Station, and the federal agency stimulated, encouraged, and directed research conducted at this local institution. During 1888 USDA assistance enabled station director William Carter Stubbs to test over seventy different varieties of foreign cane obtained overseas by United States consuls.[53] In addition, the station was selected as a site to test the diffusion process with sorghum cane, and a $10,000 grant enabled Stubbs to install a fourteen-cell diffusion battery, along with a double-effect evaporator with a capacity of four hundred square feet. New laboratory apparatuses, including a double-compensating polariscope and a Kjeldahl apparatus, were also purchased with USDA funds. Throughout the 1890s the USDA and the Louisiana Sugar Experiment Station collaborated on such varied programs as seed-cane and beet-sugar studies, soil surveys, tea cultivation, and the investigation of fiber decorticating machines.

Apparently, Commissioner Colman's decision to disengage from further direct investigations in Louisiana was based on his conviction that the USDA's objective of developing new sugar manufacturing processes was complete. He felt that it was the responsibility of private enterprise to perfect and commercialize these new findings.

Juices and Syrups," in *Report of the Commissioner of Agriculture, 1887*, 268–81; Harvey Wiley, *Record of Experiments with Sorghum in 1891*, USDA Division of Chemistry Bulletin XXXIV (Washington, D.C., 1892); Harvey Wiley, *Record of Experiments with Sorghum in 1892*, USDA Division of Chemistry Bulletin XXXVII (Washington, D.C., 1893); Charles Albert Crampton, *Record of Experiments at Des Lignes Sugar Experiment Station, Baldwin, La., during the Season of 1888*, USDA Division of Chemistry Bulletin XXII (Washington, D.C., 1889); Hubert Edson, *Record of Experiments at the Sugar Experiment Station on Calumet Plantation, Pattersonville, La.*, USDA Division of Chemistry Bulletin XXIII (Washington, D.C., 1889); Guilford Spencer, *Report of Experiments in the Manufacture of Sugar by Diffusion at Magnolia Station, La., Season of 1888–1889*, USDA Division of Chemistry Bulletin XXI (Washington, D.C., 1889).

53. Colman to Stubbs, February 15, 1888, Colman to Stubbs, July 13, 1888, Wiley to Stubbs, April 27, 1889, Wiley to Stubbs, May 29, 1889, J. M. Rusk to Stubbs, June 11, 1889, Wiley to Stubbs, January 16, 1891, all in RG 97, NA; U.S. Department of Agriculture, *Record of Experiments . . . in the Manufacture of Sugar from Sorghum*, 46–62.

In addition, USDA leadership remained convinced that the United States had the potential to become self-sufficient in sugar production if proper steps were taken to develop alternative plant sources and tropical cane cultivation in areas other than Louisiana. The USDA, which placed a renewed interest in field agriculture, established sugar experiment stations in Iowa, Nebraska, and central Florida. Scientists working at these new installations emphasized the development of improved analytical chemical methods, the application of new process reactions, and the investigation of problems dealing with plant physiology.

However, the marked shift in the USDA research program may also have been the result of Wiley's realization of his own limitations in experimental sugar manufacturing. Writing to Sieg on June 22, 1888, Wiley asserted that "I am sure there is much yet to learn in the whole matter [sugar manufacturing] and I do not know anyone more competent to go ahead than yourself. For my part I do not pretend to be a mechanical engineer and I do not see how I can do any further good in this matter. I propose hereafter to stick strictly to chemistry."[54] Between 1888 and 1892 the USDA initiated its well-known investigations of foods and food adulterants. Wiley's newly-found interest in food adulterants was not simply the response of a chemist to an emerging problem of national concern. To be sure, Wiley did see the issue of food purity as a problem involving the public interest and one that would justify future appropriations granted to the Division of Chemistry. However, Wiley felt that the solution to the problem would be found in the application of analytical chemistry, rather than in the employment of engineering techniques. In contrast with the Louisiana sugar experiments, the success of the food adulteration studies would be dependent on a body of knowledge in which he was considered an expert. Thus, he perceived that any criticism concerning his competency would be avoided in future USDA studies.

The diffusion process, which Louisiana planters had initially viewed with great enthusiasm during the 1880s, was subsequently proved to be economically unfeasible. Between 1890 and 1894, a number of planters, encouraged by a federal bounty on sugar, installed expensive diffusion batteries on their plantations. J. B. Wilkinson, Jr.,

54. Wiley to Sieg, June 22, 1888, in RG 97, NA.

of the New Orleans *Item* became a strong proponent of the diffusion process, and in 1890 he published a pamphlet extolling the advantages of the method.[55] However, his optimism, and that of other progressive planters, was quickly dampened. Additional fuel requirements and problems in the crystallization of the resultant syrup were difficult obstacles that engineering and chemical experts could not overcome.[56] For example, in tests conducted at John Pharr's Glenwild Sugar Factory during the years 1891–1893, between 220 and 255 pounds of coal were used to manufacture one ton of sugar using the diffusion process; in similar process runs during the period 1894–1896 fuel consumption was reduced to between 65 and 100 pounds per ton when an efficient mill was used. And these same investigations showed that the difference in yields using the two processes was minimal. Concurrently, new designs of mills were patented by Thomas Krajewski and Alfonso Pesant and introduced into Louisiana.[57] In response to the challenge of diffusion, the traditional three-roller horizontal mill was replaced by a more efficient combination of a two-roller mill and a nine-roller mill. Prior to extraction, the two-roller mill shredded the cane. Then either water or a dilute sugar juice was sprayed on the treated material, and the nine-roller mill completed the operation. By milling the cane and using small quantities of water in the extraction, the major complaint against the diffusion process, namely extremely dilute syrups, was avoided. By using this arrangement, both high extraction efficiencies and fuel economy were realized. In 1899 even Warmoth, the most staunch supporter of the extraction method, scrapped the diffusion cells and installed a modern mill at Magnolia.

The USDA had accomplished a great deal in Louisiana between 1884 and 1888. In particular, new machinery designs and chemical processes associated with the dynamic European beet-sugar industry were systematically introduced into Louisiana. In reality, the USDA trials had led to more questions than were answered. Yet, responsibility for research concerning the cultivation and processing of the

55. J. B. Wilkinson, Jr., *Wilkinson's Report on Diffusion and Mill Work*, 1–121.

56. "Factory Record Book, 1891–1928" (MS in John N. Pharr and Family Papers, Special Collections, Hill Memorial Library, LSU).

57. "Patents Found on Search on Two Roller Mill" (MS in A. G. Keller Papers, Special Collections, Hill Memorial Library, LSU).

cane had shifted from the USDA to local institutions during the late 1880s. During the 1890s the Louisiana Sugar Experiment Station at Audubon Park pursued many of the problems that were first recognized by Wiley and his assistants during their brief stay in Louisiana. In addition, the rapidly changing sugar industry required a new corps of well-trained chemists, and by 1890 the experiment station had begun a comprehensive program for training these scientific experts. Thus, the institutional developments that occurred in the Louisiana sugar industry during the 1890s must be evaluated within the historical context of the USDA's presence in that state during the 1880s.

8

Agricultural Research, 1885–1905

The Origins of the Louisiana Sugar Experiment Station

The primary impetus behind efforts to establish an experiment station in Louisiana can be traced to the United States Department of Agriculture and Harvey Wiley's efforts to promote science and the introduction of new technology in the state. Throughout the 1870s a number of Louisiana planters recognized that science was power, and with the end of Reconstruction a campaign was mounted to secure the services of a state chemist. The typical planter assumed that such a professional would assist agriculturists by analyzing the various soils of the state and the commercial fertilizers sold, which frequently contained considerably less nitrogen, potash, and phosphorus than claimed by the manufacturer. In addition, some naïve planters thought that upon receipt of juice or syrup samples the chemist "could at once tell what was needed to make it granulate." However, during the 1870s state legislators ignored planter requests to hire a chemist. President Johnston of Louisiana State University was willing to create a sugar laboratory at the school, but he needed financial support. The capital-poor sugar planters were unable to respond to his request for donations, however, and the issue of a state chemist seemingly died.[1]

However, these initial attempts to secure local professional exper-

1. *Louisiana Sugar Bowl* (New Iberia, La.), December 27, 1877, and May 6, 1881.

tise laid the groundwork for future developments, since they raised the question of the proper role of a state chemist and the potential benefits of chemical analyses to the Louisiana sugar industry. The issue was raised again early in 1884, but in a different institutional setting. The Louisiana Sugar Planters' Association formed a committee at its January, 1884, monthly meeting to review the ways and means of securing a chemist for the association, and throughout the year a group within the organization began to pressure the association to take a more active role in promoting technological innovation.[2] As a result, when Harvey Wiley spoke to the LSPA and met with its leaders in December of 1884, the sugar planting community was receptive to the ideas of not only hiring a chemist but also setting up an experiment station. Apparently Wiley convinced John Dymond and others of the crucial need for the services of an analytical chemist, but little more was mentioned in the LSPA minutes until May, 1885, when D. D. Colcock of the Louisiana Sugar Exchange invited William Carter Stubbs of the Alabama Experiment Station to address the association's monthly meeting.[3]

Already in New Orleans attending the World's Industrial and Cotton Centennial Exposition, Stubbs spurred the LSPA to decisive action with a speech entitled "Fertilizers and Experiment Stations." The speaker began by reviewing his twelve-year study of the effects of fertilizers on cotton yields, and he then focused on his current investigations on the cultivation of sorghum cane. While in Alabama, Stubbs could not study tropical sugar cane; however, from the perspective of sugar planters evaluating a prospective LSPA chemist, he did experiment with the next best plant—sorghum. In several tables, Stubbs summarized the results of these initial studies on the effects of nutrients on cane growth and composition. His presentation clearly reflected his abilities to devise experiments to pursue agricultural research problems systematically. The Auburn chemist demonstrated his ability to use test plats; and his analysis of the cane for the percentage of water, ash, woody fiber, aluminoids, fats, wax, and sugar demonstrated a scientific ability not unlike that of Harvey Wiley. Once Stubbs had established his competency to the

2. Louisiana Sugar Planters' Association, "Minutes of Monthly Meetings, 1877–1891," January 10, 1884, p. 103, and August 14, 1884, p. 125.

3. "The Origin of Our Sugar Experiment Station," *LP*, XXXI (1903), 266–67.

LSPA, he carefully developed a theme that had been the center of the association's discussions since 1879, namely European competition. He remarked: "The 'Beet Sugar Industry,' of Europe, has received the attention of French savants and the German experiment stations, and has brought by their assistance to such a degree of perfection as to threaten your existence as a body of cane planters. By a proper selection of seed and the application of suitable manures the sugar beet has been raised from an insignificant plant to the dignity of a formidable rival of the tropical cane."[4] Stubbs continued his argument by citing the experiments of Professor Maercker of the Halle sugar experiment station and those of French chemists, who recognized the usefulness of nitrogen, superphosphate, lime, and potash in the successful cultivation of a beet rich in sugar. Furthermore, the Alabama director asserted that the Louisiana sugar industry would remain competitive only if planters imitated their European counterparts by increasing not only the yield of cane per acre, but also the amount of sugar contained in the plants. Stubbs claimed that individual planters possessed neither the knowledge nor the time to acquire such information; only a trained agricultural chemist with access to proper laboratory facilities could hope to imitate the success of the Europeans.

In Stubbs's opinion, the sugar industry was confronted with several problems demanding immediate solutions. To begin with, the determination of manurial enrichment of the varied soils of Louisiana could be determined only in field studies. Yet the knowledge of proper dosages was insufficient; the most effective form of each nutrient also had to be established. For example, Stubbs asked whether nitrogen was most effectively applied in the cane fields in the form of sodium nitrate, ammonium nitrate, dried blood, tankage, fish scrap, vegetable nitrogen, or cottonseed meal. Once these fertilizer requirements were known, the next step was to examine the physical and chemical properties of the soil. Would surface drainage be adequate for raising sugarcane in Louisiana, or would tile drainage be necessary?

Stubbs proposed that a sugar experiment station should be more

4. Louisiana Sugar Planters' Association, *Regular Monthly Meeting, May 14, 1885* (n.p., n.d.), 5.

than test fields and analytical laboratories. He felt that manufacturing and diffusion process trials should also be initiated in a sugarhouse located at the station. Stubbs closed his message to the LSPA by proclaiming that while the obstacles were formidable, scientific methods could eventually overcome them all. "Gentlemen, remember, that whenever humanity broods over a problem, sooner or later it will be solved. Combine and concentrate your efforts in a first-class experiment station, and you will find the difficulties now encountered in sugar making, before its investigations, 'melting away, like streaks of morning light, into the infinite azure of the past.'"[5]

Within six months the association had listened to two well-prepared addresses emphasizing the threats posed by foreign competition—first Wiley's and now Stubbs's. The latter's speech motivated Dymond to stand up at the conclusion of the meeting and propose that the planters make a five-year financial commitment for the establishment of an experimental farm. Several LSPA members, including Duncan Kenner, Henry and Richard McCall, Edward Gay, T. S. Wilkinson, and Bradish Johnson, responded to Dymond's call by agreeing to support the venture.

During the summer of 1885 an advertisement calling for the creation of a planter-sponsored experiment station circulated throughout south Louisiana.[6] According to the flyer, the station would be a response to competitive scientific developments in Europe. Further, it was claimed that it could be successfully operated for the first five years by an initial subscription of $12,000, followed by four annual contributions of $6,000. However, the LSPA leadership hoped that the state would assist them by contributing a portion of these funds. They asserted that in order to secure funds for the station, lawmakers at the next state legislative session would consider a bill calling for a tax on commercial fertilizers, thus reducing the proportion required from the planters. J. Y. Gilmore, editor of the *Louisiana Sugar Bowl*, echoed current planter sentiments. "As a further incentive to subscription, I have frequently expressed the hope and belief that, after the first year, the bulk of the burden will be lifted from

5. *Ibid.*, 10.
6. "An Experimental Farm for the Promotion of the Sugar Interest," n.d. (MS in William Carter Stubbs Papers, SHC).

the planters' shoulders by the passage of an act making an appropriation to sustain the experiment station, creating our Director State Chemist, with authority to work for cotton planters and other agriculturists, empowering him to analyze all fertilizers put on the market, to protect consumers against fraud, and to enable *all* to know their component parts and relative value for the various plants."[7] Apparently, the LSPA also assumed that the Louisiana State University would fund a portion of the director's salary. In return, a second station would be created at the Baton Rouge campus.

However, considerable fund-raising difficulties were encountered, prompting the LSPA members to hire J. Y. Gilmore as a full-time employee of the experiment station planning committee. While John Dymond approached large planters for subscriptions, Gilmore was charged with contacting the smaller growers and manufacturers.[8] As a result, the proposed experimental farm was supported not only by the great sugar barons of south Louisiana, but also by a host of small-scale growers and manufacturers who made less than two hundred hogsheads of sugar during the 1884–1885 crop season. Obviously, even a few of the typical rural, tradition-bound planters were beginning to accept the idea that science had the potential to solve age-old agricultural problems.

Through Dymond's and Gilmore's efforts, ample working capital was secured by early fall, and the search for a station director was initiated. The first offer was made to the director of the North Carolina Experiment Station. When he refused, the committee approached Stubbs, and he accepted. On October 11, 1885, the LSPA executive committee reported to the association that over $10,000 had been pledged and that the Schulze plantation in Kenner, Louisiana, had been leased for five years. With that, all the preliminary steps to establish the station were accomplished.

The planters then moved to establish a separate, permanent governing body to manage the affairs of the experiment station. The Louisiana Scientific Agricultural Association (LSAA), a nonprofit corporation, was chartered on October 20, 1885, "for the purpose of

---

7. *Louisiana Sugar Bowl* (New Iberia, La.), n.d. [1885], in Stubbs Papers, SHC.

8. *Louisiana Sugar Bowl* (New Iberia, La.), n.d. [1885]; published letter, J. Y. Gilmore to John Dymond, July 7, 1885, clipping in Stubbs Papers, SHC.

developing and improving the agricultural interests and resources of Louisiana, especially the cultivation of sugar and rice by scientific and agricultural and chemical experiments and to disseminate information connected therewith."[9] Subscriptions consisting of 2,400 shares of $25 each capitalized the organization at $60,000. The powers of the LSAA were vested in a board of directors comprising forty-eight members, representing both the elite of the planter class and a few powerful New Orleans merchants. According to the corporation's charter, the LSAA's membership annually elected the board of directors; in turn, these officials chose a president, vice-president, and a six-man executive committee. Not surprisingly, the LSAA's first officers were many of the same men who had guided the LSPA since 1877. Duncan Kenner was the organization's first president, and John Dymond the first vice-president. As in the LSPA, the executive committee exhibited a balance of both merchant and planter classes, as well as Republican and Democratic party members.[10] According to the LSAA's constitution, the board of directors was charged with the election of the station's director and the approval of other staff appointees. However, in the absence of board meetings, the executive committee was empowered with governing authority. Since it appears that the board never met again after 1885, the executive committee, led by John Dymond, decided issues relating to the station throughout the 1880s.[11] Upon the death of Duncan Kenner in 1887, Henry McCall, long active in LSPA affairs, became vice-president of the LSAA, and Dymond was elected president.

Stubbs proved to be a popular and effective director of the sugar experiment station between 1885 and 1905. Born near Williamsburg, Virginia, in 1843, he began his college studies at William and Mary in 1860.[12] His studies were interrupted by a brief stint in the Confed-

9. Louisiana Sugar Planters' Association, "Minutes of Monthly Meetings, 1877–1891," October 18, 1885, p. 155; "Charter of the Louisiana Scientific Agricultural Association," 1885 (printed copy in Stubbs Papers, SHC).

10. Rudolph Sieg to Norman J. Colman, January 4, 1886, in RG 97, NA.

11. "Sugar Experiment Station," *LP*, I (1888), 177; "The Sugar Experiment Station," *LP*, III (1889), 227.

12. *National Cyclopedia of American Biography* (1935), XXIV, 208–209. Stubbs's education was traced in William Carter Stubbs to Charles Albert Browne, February 27, 1922, in Charles Albert Browne Papers, LC. Stubbs's scientific and technical con-

erate army. After capture and parole, he completed his preliminary studies at Randolph-Macon College in 1862 and again served in the Confederate cavalry until the surrender at Appomattox. In 1865 Stubbs enrolled at the University of Virginia, where he completed the master's course in chemistry and geology in 1867. He then studied analytical chemistry under Professor John William Mallet for one year.[13] An Englishman, Mallet received his doctorate under Friedrich Wöhler at Göttingen in 1853. He then came to America where he received an appointment at Amherst College. In 1855 he became a chemist for the Alabama State Geological Survey, and a year later was hired as professor of chemistry at the University of Alabama. During the Civil War he headed a Confederate army ordnance laboratory, and then worked briefly at the Louisiana State University Medical School before assuming his position at the University of Virginia. His 1862 treatise *Cotton: The Chemical, Geological, and Meteorological Conditions Involved in its Successful Cultivation* had considerable influence upon Stubbs's later studies. Above all, Mallet impressed upon Stubbs that chemistry was a quantitative rather than a qualitative discipline. Between 1869 and 1872 Stubbs served as professor of chemistry at East Alabama College, and in 1872 he was appointed professor of chemistry at Alabama Agricultural and Mechanical College at Auburn. Stubbs gained additional responsibilities in 1877, when he assumed the position of Alabama state chemist.

Stubbs's work in Alabama focused on agricultural fertilizer analysis and the testing of several crop varieties. As state chemist, he ana-

---

tributions can be best reviewed by examining his many addresses and papers appearing in *The Louisiana Planter and Sugar Manufacturer.* In addition, his monographs include: *Sugar Cane: A Treatise on the History, Botany and Agriculture of Sugar Cane, and the Chemistry and Manufacture of its Juices into Sugar and Other Products* (New Orleans, 1897); *Sugar Cane* (Boston, 1903); and with Herbert Myrick, *The American Sugar Industry: A Practical Manual on the Production of Sugar Beets and Sugar Cane, and on the Manufacture of Sugar Therefrom* (New York, 1899).

13. *National Cyclopedia of American Biography*, XIII, 55. John William Mallet, *Chemistry Applied to the Arts: A Lecture Delivered Before the University of Virginia, May 20, 1868* (Lynchburg, Va., 1868); Mallet, *Cotton: The Chemical, Geological and Meteorological Conditions Involved in its Successful Cultivation* (London, 1862).

lyzed numerous water and iron ore samples. Beginning in 1881, Stubbs examined the effects of nitrogen fertilizers on cotton, peas, potatoes, beans, peanuts, legumes, and grasses. Later Stubbs would transfer the methods and techniques learned at Auburn to new agricultural problems in Louisiana. Throughout his career Stubbs's research focused on field studies. Most significantly, his conception of agricultural chemistry was rooted in traditional American ideas and practices common between 1860 and 1880. His scientific activities in Louisiana were strongly influenced by Samuel W. Johnson's popular treatises *How Crops Grow* and *How Crops Feed*.[14] Stubbs employed Johnson's views not only in formulating the research program at the Louisiana Sugar Experiment Station, but also in teaching scientific agriculture to students.[15] Like Johnson, Stubbs's understanding of plant growth was based on the analysis of inorganic ash residues and organic compounds, an approach which was the result of chemical and agricultural investigations first proposed by Liebig in the early 1840s and employed extensively by European and American chemists primarily during the late 1840s and 1850s. Stubbs correlated plant growth stages with the relative quantities of inorganic acidic and basic salts found when various portions of the plant were ashed. Furthermore, he estimated the volatile components of plants (nitrogen, carbon, hydrogen, oxygen, phosphorus, and sulfur) and extracted such loosely defined organic constituents as fibers, fats, and albuminoids.

For Stubbs, inorganic ash analysis provided the data necessary for a systematic evaluation of plant growth. During the 1870s and 1880s it was assumed that potash, soda, lime, magnesia, oxide of iron, chlorine, sulfuric acid, phosphoric acid, silicic acid, and carbonic acid

14. William Carter Stubbs, "Notes on Chemical Analysis" (MS in William Carter Stubbs Papers, Swem Library, College of William and Mary, Williamsburg, Va.); Samuel W. Johnson, *How Crops Grow: A Treatise on the Chemical Composition, Structure and Life of the Plant, For all Students of Agriculture* (New York, 1887); Samuel W. Johnson, *How Crops Feed: A Treatise on the Atmosphere and the Soil as Related to the Nutrition of Agricultural Plants, with Illustrations* (New York, 1882). For information on Johnson, see Margaret Rossiter, *The Emergence of Agricultural Science: Justus Liebig and the Americans* (New Haven, 1975), 127–48.

15. See Stubbs's junior- and senior-year examinations at the Audubon Sugar School in *LP,* XIV (1895), 403–405.

were indispensable for the life and growth of all plants. Plant ash varied in composition according to the botanical nature of the plant, its relative stage of growth, vigor, nourishment, and soil. As a result, Stubbs evaluated his field work not only by observing and calculating crop yields, but also by performing laboratory analyses of the constituents in plant ash. However, Stubbs's research overlooked the organic reaction processes taking place within living plants, such as the conversion of one form of organic nitrogen compound into another. His conception of agricultural chemistry failed to incorporate ideas related to the development of organic structural theory and advances in carbohydrate chemistry, two areas of research that had dominated the investigations of organic chemists during the last half of the nineteenth century.

Organization and Research, 1885–1890

On October 5, 1885, Stubbs began to cultivate numerous sugarcane test plats at the LSAA's newly leased plantation in Kenner, Louisiana.[16] The director's long-range objective was to determine the optimum conditions for the cultivation of high yields of sugar-rich cane. He studied specific fertilizer requirements, soil characteristics, proper planting practices, and he examined the possibility of cultivating foreign plant varieties. Throughout his career, Stubbs was intrigued by one major research problem—the physical and chemical dynamics of plant growth. Although he was subsequently pressured into investigating other problems of immediate concern to planters and manufacturers, he consistently attempted to orient seemingly disparate station studies around this focal point.

Stubbs's understanding of the role of soils influenced his views

16. On the history of nineteenth-century experiment stations, see Charles E. Rosenberg, "Science, Technology and Economic Growth: The Case of the Agricultural Experiment Station Scientist, 1875–1914," *Agricultural History*, XLIV (1971), 1–20; Charles E. Rosenberg, "The Adams Act: Politics and the Cause of Scientific Research," *Agricultural History*, XXXVIII (1964), 3–21; Alfred Charles True, *A History of Agricultural Experimentation and Research in the United States, 1607–1925*, USDA Miscellaneous Publication CCLI (Washington, D.C., 1937); Jane M. Porter, "Experiment Stations in the South, 1877–1940," *Agricultural History*, LIII (1979), 84–100; C. W. Stewart, "A Brief Discussion of the Audubon Sugar Factory" (Typescript in Special Collections, Hill Memorial Library, LSU).

concerning the action of fertilizers on plant growth. Unlike George Ville, the French agricultural scientist who had assumed that soil was merely an inert receptacle for fertilizers, Stubbs viewed soil as a dynamic mixture of organic and inorganic substances possessing a natural strength, indeed a vitality of its own. He felt that soils were capable of furnishing at least a portion of the nutrients necessary for proper plant growth. Yet Stubbs's ideas on plant nutrition were not original; they were based on the concepts propounded by S. W. Johnson, J. B. Lawes, and W. O. Atwater. In addition, in one of his first bulletins Stubbs cited the popular 1880s treatise by Joseph Harris entitled *Talks on Manure*. For specific knowledge on sugarcane fertilizers, Stubbs initially drew on the results of field trials at St. Denis to guide him in his own investigations.

The newly appointed director systematically varied the proportions of cottonseed meal, acid phosphate, kainite, gypsum, cotton hull ashes, tankage, and stable manure in different types of soils. Stubbs's test plats were organized in a way that he could determine the natural strength of the soil, the effect of nitrogen, phosphorus, and potash alone on cane growth, the result of combining any two of these nutrients, and the yield from mixing all three elements in several ratios. In the spring of 1886 these fertilizer experiments were continued and field trials were initiated to test the merits of various commercial fertilizers and varieties of sorghum, oats, corn, and rice.[17] These field experiments also included germination studies, which Stubbs hoped would settle one important question dividing the Louisiana planters of the 1880s—whether tops, middles, or butts of cane served as the best seed. Cane stalks consist of joints, connected by nodes. Each node contains a bud or "eye," which sprouts when planted. The stalk was cut into sections having two or three buds each. During sowing, these cuttings were dropped into the furrows by hand. Cane tops, the most immature portion of the plant, were the most economical seed available to planters. However, Stubbs set out to systematically determine whether the resulting cane was inferior to that of a plant having mature butts as seed.

The work in the fields was only one aspect of the station's activi-

17. Louisiana Sugar Experiment Station, *Bulletin No. 2 . . . January 20, 1886* (N.p., n.d.), 8, 17.

ties. The station's laboratory for the analysis of patrons' fertilizer and soil samples was completed in the spring of 1886. It consisted of a furnace room, a large wet-laboratory, a balance room, a polariscope room, and a small storage area. The furnace room contained a two-horsepower boiler, water and steam baths, a drying oven, a still, and several types of furnaces. Laboratory operations—precipitations, volumetric measurements, and the preparation of solutions—took place in the so-called work room, while final polariscopic and weight determinations were conducted in adjoining rooms.[18]

However, in the fall of 1886 Stubbs's ambitious program was jeopardized by a serious shortage of operating funds. In the station's first annual report, the treasurer of the LSAA, W. B. Schmidt, reported that expenditures exceeded income by over three thousand dollars. This shortfall was largely the result of numerous subscribers failing to submit pledged monies. Stubbs and his assistants had not been paid in three months, and the rents on the plantation grounds and sugarhouse were overdue. Schmidt recommended that "it is highly important to provide some means to meet the deficiency by acquiring more members to join the association; and I would also recommend to curtail the running expenses whenever possible."[19]

Even before the financial report was issued, Stubbs had anticipated the shortfall and had already begun to seek alternative sources for funding his own salary and some of the station's expenses. In response to pressure from Louisiana farmers, but independent of Stubbs and the sugar planters, the Louisiana State Board of Agriculture had established a general experiment station on Louisiana State University property in February of 1886.[20] Even though the station was considered to be a department of the university, it was legally under the direction and control of the Board of Agriculture. Five months later Stubbs signed a contract with the commissioner of agriculture in which he became the official state chemist, director of the Baton Rouge Experiment Station, and professor of agriculture at Louisiana

18. Louisiana Sugar Experiment Station, *Bulletin No. 3 . . . April 1, 1886* (N.p., n.d.).

19. *Ibid.*, 4. For an example of the report issued by the station, see reports to H. C. Warmoth, March 22, 1887, in Warmoth Papers, Reel 19, SHC.

20. W. B. Schmidt to the President and Members of the Louisiana Scientific Agricultural Association, October 1, 1886, in Stubbs Papers, Folder 1, SHC.

State University for a salary of $1,000 annually. In turn, the commissioner of agriculture agreed with the LSAA that the sugar planters would remain in control of the Kenner sugar experiment station, provided that they would forward station results to the state for publication. Furthermore, since the LSAA would permit Stubbs to use the station's laboratory for state regulatory fertilizer analysis, the Bureau of Agriculture then allocated one-half of the state's fertilizer fund for use at the Kenner sugar experiment station.

Stubbs, the LSAA, and the commissioner of agriculture had entered into a complex set of contracts that became even more complicated after the passage of the Hatch Act in 1887, which allocated $15,000 for experiment stations in each state. This federal law also stipulated that experiment stations established after the passage of this bill would be managed by the state's land-grant university. However, since the Baton Rouge and Louisiana sugar experiment stations were already under the joint control of the Louisiana State Board of Agriculture and the LSAA, a strong effort was made by planters and certain politicians in Louisiana, possibly under Stubbs's leadership, to maintain the separation of the experiment stations from the university.[21] In the spring of 1887, the Louisiana State University Board of Supervisors formally agreed with this arrangement. The Board of Agriculture consisted of the governor, the vice-president of the university's board of supervisors, the commissioner of agriculture and immigration, the president of the university, the director of experiment stations, and six members appointed by the governor, one from each congressional district.[22] Although not legally tied to the state university, the Baton Rouge Experiment Station made provisions to allow university students adequate opportunities to observe and study the experiments conducted there.

During 1887 the ultimate number and locations of these research centers also proved to be a point of controversy. Colonel H. M. Favrot of the Louisiana State University Board of Supervisors wanted the

21. W. C. Stubbs to Thomas D. Boyd, June 12, 1900, in William C. Stubbs Papers, LSU.

22. Walter L. Fleming, *Louisiana State University, 1860–1896* (Baton Rouge, La., 1936), 419–21, 478–82; Charles A. Browne to W. E. Cross, April 24, 1923, in Browne Papers, Box 3, LC.

Hatch Act to fund only one strong station in Baton Rouge. Farmers in north Louisiana, however, were demanding that a station be funded in their area as well, since soil conditions, crops, and animal breeding problems were completely different from those found in south Louisiana. Ultimately it was decided that three independent stations would divide the Hatch Act funds equally. Thus, after 1887, Stubbs directed three experiment stations, located in Baton Rouge, Kenner, and Calhoun, Louisiana. Two of the three stations were controlled by the Louisiana State Board of Agriculture, while the LSAA planters maintained control of the sugar experiment station at Kenner through a complex legal arrangement.

The decisions made in 1887 had a lasting impact on the nature of work done at experiment stations in Louisiana and the politics associated with these institutions. Although Louisiana experiment stations were well funded relative to those of other states, the decentralization of agricultural research diluted the available financial and scientific resources of the state. Further, the issue of control arose again between 1888 and 1910, when a struggle over station management ensued between the Board of Agriculture and Louisiana State University. The fact that not all were happy with the arrangement was apparent in 1888 when William Garig, vice-president of the Louisiana State University Board of Supervisors, argued that the university should have legal jurisdiction over the stations.

As a consequence of these legal maneuvers, Stubbs had secured financial support for himself and for the sugar experiment station. Although he was professor of agriculture at the university and director of all three stations, Stubbs's first priority was managing the Kenner station near New Orleans. In spite of extensive commitments in Baton Rouge, he resided in New Orleans and spent the majority of his time working on sugar-related studies. Thus, by 1888 Stubbs was finally in a position to expand the station's research staff and broaden the scope of its investigations.

While Hatch Act funds were important to the success of Stubbs's research program at the sugar experiment station, the director also depended on three other sources—the United States Department of Agriculture, private industry, and the sugar planters. Between 1888 and 1900 the USDA supplied the Louisiana Sugar Experiment Station with the money and materials necessary to broaden the scope of its research program. As a result, Stubbs and his staff were able to investigate new sugarcane varieties, to examine the diffusion process, and to conduct a statewide soil-mapping survey. Also, industrial concerns and private interests made significant contributions to the station's physical plant. Stubbs's task of setting up a modern sugarhouse in 1887 was facilitated by donations of machinery from equipment manufacturers like the Edwards and Haubtman Company and Whitney Iron Works of New Orleans; they realized that the station's endorsement would lead to a competitive edge in the local marketplace. For example, the Yaryan Manufacturing Company of Ohio supplied Stubbs with a vacuum distilling apparatus.[23] Perhaps anxious to display their products to the numerous planters who would undoubtedly inspect the station's manufacturing equipment, local foundries and manufacturers' representatives donated mixers and settling tanks, shafting, pulleys, and filter presses.

Although the federal government and local foundries contributed to the station's funding and thereby influenced its research programs, the LSPA remained its chief patron and the dominant force in station affairs until the turn of the century. In fact, by 1890 the LSPA executive committee had displaced the LSAA as the body to which Stubbs reported annually. In 1894 the LSPA contributed $10,000 toward station expenses, compared to $5,000 received from the Hatch Act and $2,000 from the Louisiana State Bureau of Agriculture. Although LSPA support varied from year to year—at times totaling less than $4,000, and on other occasions more than $10,000—the planters, and not the state, ultimately controlled station policy.

From the beginning, Stubbs deliberately involved the planters in station activities. Before the procurement of Hatch Act funds, Stubbs attempted to promote systematic investigations by station patrons themselves. Because of the marked variances in soil characteristics from one area to another in Louisiana, he felt that "there is . . . a necessity for individual experimentation. Every farmer or planter should conduct yearly a series of experiments upon his crops for himself, results of which, if rightly obtained and interpreted, will be

23. Alfred Charles True and V. A. Clarke, *The Agricultural Experiment Stations in the United States* (Washington, D.C., 1900), 237.

of incalculable benefit. It would ultimately redound to his pecuniary benefit, besides cultivating his powers of observation and reflection, and making him a philosopher and student as well as a farmer."[24] To insure that planters followed proper agricultural procedures, Stubbs then outlined the steps that he hoped planters would take. He proposed that each planter should lay out ten 1-acre test plats and keep a record of the time of planting, subsequent workings, and meteorological conditions. At the harvest, a portion of the cane from each plat would be weighed, percent yield calculated, and the stalks then sent to the station laboratory for chemical analysis.

These voluntary planter efforts were one strategy employed by Stubbs to involve the patrons with the station. Questionnaires also served to link planters with the station staff. Planters reciprocated by donating cane varieties and seedlings for experimental field trials. Perhaps the most significant bond was developed through Stubbs's frequent contributions to the LSPA monthly meetings. There, he presented his current research findings and future plans in scientific language simple enough for the planters to understand. For example, during 1888 Stubbs spoke on such diverse topics as "The Benefits of Humus to the Soil," "Commercial Fertilizers," and "Sugar House Chemistry."

Following increased support, Stubbs expanded the station staff to seven professional employees and delegated specific responsibilities to his subordinates.[25] Maurice Bird, a chemist trained at the University of Virginia, was appointed assistant director in 1888 and was placed in charge of the routine fertilizer and pesticide analyses. While two assistant chemists worked with Bird in the laboratory, John P. Baldwin of St. Mary Parish was placed in charge of the sugarhouse, and a fourth chemist was given responsibility for operation of the diffusion apparatus. A farm manager supervised the day-to-day field activities, and a secretary/bookkeeper kept track of the finances.[26] While LSPA support varied from $4,000 to $12,000 and state

24. William C. Stubbs, *Sugar Cane: Sugar House and Laboratory Experiments—1886*, Louisiana Sugar Experiment Station Bulletin X (Baton Rouge, 1886).

25. Louisiana Sugar Experiment Station, *Bulletin No. 2*, 15.

26. U.S. Department of Agriculture, Office of Experiment Stations, *Organization of the Agricultural Experiment Stations in the United States: February 1889* (Washington, D.C., 1889).

monies averaged about $2,000 per year, proceeds from the sale of farm products amounted to roughly $1,000 annually. Also, the USDA proved to be at times a generous patron to the sugar experiment station, as in 1890 when Stubbs received $10,000 for sorghum studies. By 1889 the Louisiana Sugar Experiment Station had emerged with a formal organizational structure. During the next year the institution would move to a more favorable location and assume new functions.

In 1888 LSPA leaders initiated plans to move the station from its leased location near Kenner, Louisiana, to a site in uptown New Orleans. David Calder and J. C. Morris made arrangements with the commissioner of Audubon Park to build a new station within easy commuting distance from the downtown business district.[27] The LSPA membership spent nearly $100,000 to equip the newly erected buildings and laboratories at the park. Thirty acres were devoted to flowers and trees, and the large Horticultural Hall was located on this portion of the station's grounds. This glass building, measuring six hundred by two hundred feet, was stocked with a collection of tropical fruits (palms, bananas, coffee, pineapple), along with foreign varieties of sugarcane. On another fifty acres a large iron sugarhouse, an iron barn, a stable, the stabler's house, the station director's residence, and a two-story laboratory were built. The laboratory contained a library, an office, polariscope and balance rooms, an assembly room, and a large workroom. The second floor contained an agricultural museum, as well as botanical and microscopical laboratories.

The surrounding fields included test plats devoted not only to sugarcane, but also to small grains, grasses, and clovers, fruit, truck gardens, fiber plants, rice, corn, and peas. Fertilizer studies were also conducted on numerous varieties of cotton, sorghum, potatoes, beets, lettuce, radishes, peppers, cabbage, kohlrabi, tomatoes, beans, squash, carrots, melons, turnips, peanuts, and other crops. Nearby, an orchard contained orange, fig, plum, and pear trees.

After 1890 the Audubon Park Station evolved into an agricultural

27. W. C. Stubbs to W. O. Atwater, January 21, 1891, and Stubbs to Charles W. Dabney, February 12, 1895, all in Record Group 164, Records of the Office of Experiment Stations, Louisiana File, NA.

research center with field investigations that were often either marginally connected or totally unrelated to sugar growing and manufacturing. This broadening of scope was in part due to a new responsibility imposed by the state government on the station—the attraction of potential immigrants to Louisiana. The station grounds and buildings imparted a powerful visual stimulus to visitors—one by which the promises of modern agricultural science and the rewards of farming were transformed into realities. The Louisiana State Board of Agriculture had as one of its responsibilities the recruitment of white labor into Louisiana, and the station served the purpose of exhibiting the various types of crops that the enterprising and intelligent farmer could easily cultivate. Members of the Louisiana Sugar Planters' Association were also served indirectly by such publicity, since an influx of small landowners or tenants in the state could contribute to the industry's development.[28]

Also, the station's activities included testing and endorsing promising inventions related to sugar manufacture. For example, a tower, $20 \times 20 \times 60$ feet, was erected in 1890 beside the experimental fields by a Dr. Murrell of Philadelphia. This tower contained the inventor's "Cyclone Process" for the evaporation of sugar juices. Large groups of planters often came to the station to witness the evaluation of such experimental apparatuses, as in 1892, when the USDA and station staff held a public trial of ramie decorticating machines.[29] Ramie was thought to be a possible source of silklike fibers, and though the 1892 trials ended in failure, a London textile syndicate approached Stubbs in hopes that an evaluation of improved machinery could take place in 1894. Stubbs and two other agricultural scientists acted as judges. They concluded that the mechanical im-

28. A. A. Maginis to W. C. Stubbs, December 1, 1888, in Stubbs Papers, Folder 124, College of William and Mary.

29. J. Chronegh Morrison, *Louisiana and its Resources: The State of the Future: An Official Guide for Capitalists, Manufacturers, Agriculturists and the Emigrating Masses* (N.p. [1886?]; Louisiana Bureau of Immigration, *An Invitation to Immigrants. Louisiana: Its Products, Soil, and Climate as Shown by Northern and Western Men Who Now Reside in this State* (Baton Rouge, 1894). See also W. C. Stubbs, *A Hand-Book of Louisiana . . . Giving . . . Geographical and Agricultural Features* (New Orleans, 1895), 31–41.

provements were inadequate for economical operations. Another trial that took place in 1894 was the investigation of the effects of electric current on clarification. In November, a New York inventor installed patented aluminum electrodes in the clarifier tanks at the sugar experiment station. However, no visible changes in the impure juices took place. Yet public trials continued at the station, particularly after 1900, when an intensified interest in cane loaders emerged as a consequence of the unskilled labor shortage. Hundreds of planters attended these well-publicized demonstrations.[30]

Like the Louisiana State Department of Agriculture, the LSPA was also interested in broadening the sugar experiment station's responsibilities. The planters felt that the station was well equipped and staffed to train young men to become sugar experts—scientists capable of handling the agricultural, chemical, and engineering problems encountered in the efficient management of a central sugar factory. In 1891 the Audubon Sugar School was established on the station grounds, with its faculty consisting of the station staff.

Within five years Stubbs had organized a research experiment station and had carried out a series of studies that were centered around field experiments and correlative analytical chemical data. The competitive influences of state and federal funds and the local sugar manufacturing interest somewhat diffused this investigative program. Yet the director concentrated his efforts on agriculture, since he felt that the Louisiana sugar industry was primarily an agricultural activity. However, events taking place in Washington, namely the passage of federal bounty legislation, ultimately restructured the business. During the 1890s the sugar industry separated into two interest groups—growers and manufacturers. As a result, new demands were placed on the station to establish a more scientific and technical basis for sugarhouse reactions and processes. The staff of the Louisiana Sugar Experiment Station responded by conducting investigations that satisfied the needs of a planter and manufacturing class more interested in technical developments in the sugarhouse

30. William C. Stubbs, *Ramie— . . . Uses, History, . . . With a Report of Committee on the Recent Public Trial of Ramie Machines, at Audubon Park, New Orleans, January 1895,* Louisiana Experiment Station Bulletin XXXII (Baton Rouge, 1895); Sitterson, *Sugar Country,* 278. Also see *LP,* XXXV (1905), 26–28.

than advances in agricultural chemistry. However, these production investigations were pursued in such a manner that complex problems in plant growth were studied simultaneously. A new branch of carbohydrate chemistry was explored at the Audubon Park station, and for a brief time it became an internationally recognized research center.

The "neat best looking fellow" at the Audubon Sugar School, 1894. The school, established in 1891, trained students in both chemistry and mechanical engineering.
Special Collections, Hill Memorial Library, Louisiana State University

Louisiana Sugar Experiment Station, Audubon Park, 1899. Erected in 1890 on the present site of the New Orleans Audubon Zoo, the experiment station laboratory was a focal point of nineteenth-century scientific investigations related to the manufacture of sugar.
Special Collections, Hill Memorial Library, Louisiana State University

Experimental sugarhouse, Louisiana Sugar Experiment Station, Audubon Park. Under the direction of William C. Stubbs, cane from nearby test plats was processed using a variety of methods. During grinding season, students enrolled in the Audubon Sugar School gained practical experience in the manufacture of sugar.
Special Collections, Hill Memorial Library, Louisiana State University

Experimental evaporator and pan, Audubon Sugar School, 1902. Designed and patented by Norbert Rillieux during the 1840s, the multiple-effect evaporator greatly decreased fuel costs and by the 1880s was widely accepted in Louisiana.
Special Collections, Hill Memorial Library, Louisiana State University

In the sugarhouse, Audubon Park, 1909. At the top of the photograph are the multiple-effect evaporators; on the level below are the centrifugals used in separating crystalline sugar from molasses.
Special Collections, Hill Memorial Library, Louisiana State University

"Clarifying Sugar, Titration at the Clarifiers," 1907. Beginning in the 1880s, volumetric analysis was increasingly used to control certain steps in the manufacture of sugar.

Special Collections, Hill Memorial Library, Louisiana State University

9

Carbohydrate Chemistry at Audubon Park

Once the experimental fields were planted and the laboratory and sugarhouse apparatuses were in place at Audubon Park, Stubbs began to consider seriously the station's role in investigating sugar manufacturing processes. It is not clear whether the director was merely accommodating the demands of his planter patrons, who were debating the merits of the diffusion process, or whether he sincerely considered manufacturing to fall within the domain of the station's activities. At any rate, Stubbs arranged manufacturing studies so that data from large-scale sugar production tests would also yield information relative to the formation of complex carbohydrates and invert sugars in the cane. Sugarhouse data were used to understand not only the process of maturation of sugarcane, but also the relationship between chemical composition and growth. The conversion of glucose into sucrose, and the complex interaction of gums, pectose sugars, and albuminoids posed manufacturing problems that ultimately led staff scientists to answer questions concerning the growth of the cane. Thus, fermentation, an ever-present threat to efficient processing, was closely tied to the presence of albuminoids. An understanding of these substances and their associated reactions revealed the nature of plant growth.[1]

1. William C. Stubbs, *Sugar Cane: Sugar House and Laboratory Experiments—1886*, 12.

During the 1890s, experiments in the fields, laboratory, and sugar-house at Audubon Park were unified by concentrating on the nature of nitrogen compounds in plants. And it was in the laboratory that Stubbs's overall concern with plant growth could be linked with dis-tantly related reactions taking place in the sugarhouse. Research into the complexities of nitrogen compounds was facilitated by the adoption of the Kjeldahl method of nitrogen analysis, which proved to be tailored to the needs of the experiment station.

Until the mid-1880s, nitrogen analysis was tedious and compli-cated; the chemist usually followed either the well-known Dumas or the Will-Varentrapp method.[2] In the Will-Varentrapp procedure, first published in 1841, the sample and soda lime were placed together in one end of a drawn-out tube, while the other end was connected to an acid-filled receiver. The contents were carefully heated, and nitro-gen subsequently evolved as ammonia or a related base. Upon com-pletion of the reaction, residual gases were drawn into the receiver with the assistance of an aspirator, and nitrogen could be deter-mined by back-titrating with a standard base.

John Kjeldahl, a Danish agricultural chemist, introduced a conve-nient and rapid method of nitrogen analysis in 1883.[3] Kjeldahl found that ammonia could be accurately determined by first treating the sample with boiling, concentrated sulfuric acid, a process that con-verted the nitrogen contained in organic compounds into ammonium sulfate. Potassium sulfate was usually added to the reaction mixture to raise its boiling point, along with various metallic salts to hasten the decomposition of organic materials. After the digestion step, an excess of sodium hydroxide was added to the treated sample, and the resulting ammonia was distilled into a receiver containing an excess of standard acid. Finally, a titration using a standard base concluded the assay.[4] Stubbs employed this convenient analytical method as early as 1886 to answer manufacturing inquiries from the sugar-house, as well as questions concerning soil, fertilizer, and cane com-positions in the test fields.

2. Ralph E. Oesper, "Kjeldahl and the Determination of Nitrogen," *Journal of Chemical Education*, XI (1934), 457–62.

3. *Dictionary of Scientific Biography*, VIII, 393–94.

4. Herbert A. Laitinen and Walter E. Harris, *Chemical Analysis* (New York, 1975), 115.

Stubbs, however, was unfamiliar with the basics of carbohydrate chemistry, and between 1890 and 1905 he relied on the research efforts of a group of chemists who had been trained in the fundamentals of organic structural chemistry. Thus, during the 1890s a new type of agricultural chemistry, distinct from that previously employed by Stubbs, was practiced at the Audubon Park experiment station. Investigators focused on problems related to plant biochemistry, and they often employed new ideas and techniques originated in European laboratories.

A number of chemists had found employment at the station between 1888 and 1905. Some, like T. H. Jones, Maurice Bird, and W. Wipprecht, had been trained at southern universities.[5] Wipprecht, however, was also educated at Göttingen, and he proved to be the first of a large number of Bernard Tollens' former students, including Horace E. Horton, Charles A. Browne, Fritz Zerban, Peter Yoder, and William E. Cross, to apply a theoretical knowledge of sugar chemistry to the solution of practical problems at the Louisiana Sugar Experiment Station.[6]

Bernard Tollens, who had received his training under Rudolf Fittig at Göttingen, Emil Erlenmeyer at Heidelberg, and Charles Adolphe Wurtz at Paris, had succeeded Wilhelm Wicke in 1873 as professor and director of the laboratory at Göttingen's Agricultural Institute. Although every student of elementary organic chemistry is familiar with Tollens because of the test reagent that bears his name, historians of science have neglected to study his career and scientific contributions. Working in the shadow of the great Emil Fischer, Tollens was a crucial figure in the development of organic chemistry in Germany and had important connections with the industrial community. Until 1911 Tollens and his students concentrated their efforts on determining the exact chemical nature of the unknown carbohydrates contained in the crude fibers and nitrogen-free extracts of crops. Emil Fischer's work on the structure of carbohydrates enabled Tollens to understand basic carbohydrate reactions like hy-

5. *LP*, I (1888), 107.

6. Charles A. Browne, "Bernard Tollens (1841–1918) and Some American Students of His School of Agricultural Chemistry," *Journal of Chemical Education*, XIX (1942), 253–59; Aaron J. Ihde, *The Development of Modern Chemistry* (New York, 1964), 344–45.

drolysis, hydrazone formation, and fermentation.[7] With this knowledge, Tollens successfully separated and analyzed many carbohydrate compounds normally found in plant materials, including hexosans, pentosans, and methyl pentosans. He also conducted studies on enzyme reactions and the process of fermentation. Similar studies became the focus of the Louisiana Sugar Experiment Station after 1900, when several of Tollens' former students applied their knowledge of carbohydrate and structural chemistry to the solution of practical sugar manufacturing problems.

Although a majority of the chemists employed at the Louisiana Sugar Experiment Station between 1889 and 1910 were former students of Tollens', graduates of Harvard, Johns Hopkins University, and Zürich Polytechnic Institute also made valuable research contributions. Two station chemists, H. E. L. Horton and J. T. Crawley, were educated at Harvard.[8] Crawley had studied agricultural and physiological chemistry at Harvard under Walter Maxwell, and in 1893 the student was replaced at the Audubon Sugar School by his mentor. In 1893 Stubbs's research efforts benefited from the addition of two chemists with doctoral degrees to the station's staff, Jasper L. Beeson and Walter Maxwell. While manufacturing problems and questions on plant growth remained central, Beeson and Maxwell used their own talents to gradually redirect the nature of station investigations.

J. L. Beeson was born in Alabama in 1867.[9] After graduating from the University of Alabama in 1889, he was appointed instructor of physics at the university, and a year later he found employment as a chemist for the Alabama Geological Survey. In 1891 he entered Johns Hopkins University, where he earned his doctorate in 1893. He then joined the staff at the Louisiana Sugar Experiment Station, where he conducted research and taught chemistry to students of the station's sugar school.

7. B. Tollens, *Kurzes Handbuch der Kohlenhydrate* (Breslau, 1895).

8. H. E. L. Horton, "Some Notes on the Determination of Sugars with Fehling's Solution," *Journal of Analytical Chemistry*, IV (1890), 370–81; J. T. Crawley, "A Simplified Fat-Extracting Apparatus," *American Chemical Journal*, XI (1889), 507–508.

9. Jasper Luther Beeson, "A Study of the Action of Certain Diazo-Compounds on Methyl and Ethyl Alcohols Under Varying Conditions" (Ph.D. dissertation, Johns Hopkins University, 1893).

Beeson's research investigated the link between problems in sugar manufacture and the processes of plant growth and nutrition. The most controversial topic among Louisiana planters and sugar manufacturers during the early 1890s was whether diffusion or milling was more economical. Since it had been noticed that juices obtained from two successive milling operations were completely different in color, it was postulated that the material contained in the cane internodes was different from that expressed from the nodes during the second milling. Beeson found that the substances extracted from the nodes were highly colored, giving a heavy precipitate when mixed with a solution of subacetate of lead and coagulating when heated to the boiling point. However, extracts from the internodes were clear, light in color, producing little coagulation upon heating, and containing more albuminoids and reducing sugars than the internodes. In response to these findings, Beeson hypothesized that the physiological function of cane nodes was similar to that of seeds in flowering plants—to store food for the sustenance of the young plant during the initial period before it had taken root sufficiently to draw nutrients from the earth.[10]

In Beeson's experiments on pedigreeing cane, he chemically analyzed sugarcane grown from seed tops, middles, and butts and found no significant differences. However, he discovered the presence of less sugar and more solids in the tops (albuminoids, amide nitrogen, and ash). Therefore Beeson suggested that the tops be used for seeds, cutting down to the first joint that had cast its leaves and sending the remaining sugar-rich portion to the mills for processing.

Similarly, Walter Maxwell also studied fundamental problems in plant biochemistry that had their origin in manufacturing. Maxwell, who was born in Britain in 1854 and had studied at the science department at South Kensington, later studied with Professor Ernst Schulze at Zürich, where he investigated the chemistry of plant cell membranes and the constituents of leguminous seeds.[11] In 1888

10. J. L. Beeson, "A Study of the Constituents of Nodes and Internodes," "The Estimation of Crude Fibre in Sugar Cane," "Pedigreeing of Cane—'Tops from Tops,'" and "Effects of Fertilizers Upon Sugar Cane," in The Chemistry of Sugar Cane and Its Products, Louisiana Sugar Experiment Station Bulletin, XXXVIII (Baton Rouge, 1895), 1341–71.

11. C. A. Browne, "Dr. Walter Maxwell," Facts About Sugar, XXVII (1932), 24. One

Maxwell went to Harvard University, where he conducted a course in physiological chemistry. Between 1889 and 1893 he served as the USDA assistant chemist supervising the agency's beet-sugar experiments at Schuyler, Nebraska. Maxwell accepted a position at Audubon Park in 1893, and he remained in Louisiana until appointed in 1895 to the directorship of the Hawaii Sugar Experiment Station.

Maxwell's work at the Louisiana Sugar Experiment Station was significantly influenced by his previous training with plant biochemist Ernst Schulze. Born in 1840, Schulze had studied with Friedrich Wöhler and Heinrich Limpricht at Göttingen and later under Robert Wilhelm Bunsen in Heidelberg. In 1872 he was appointed professor of agricultural chemistry at the Zürich Polytechnic Institute, where he remained until his death in 1912.[12] Schulze, who

---

can readily ascertain the influence of Ernst Schulze on Maxwell by examining the latter's early publications. See W. Maxwell, "On the Solubility of the Constituents of Seeds in Prepared Solutions of Ptyalin, Pepsin, and Trypsin," *American Chemical Journal*, XI (1889), 354–57; "On the Presence of Sugar-Yielding Insoluble Carbohydrates Present in the Seeds of Legumes," *ibid.*, XII (1890), 265–69; "On the Methods of Estimation of Fatty Bodies in Vegetable Organisms," *ibid.*, XIII (1891), 13–16; "On the Behavior of the Fatty Bodies, and the Role of the Lecithines, During Normal Germination," *ibid.*, XIII (1891), 16–24. Maxwell's interest in nitrogen compounds is reflected in "On the Nitrogenous Bases Present in the Cotton Seed," *ibid.*, XIII (1891), 469–71. His early studies on germination were published as "Movement of the Element Phosphorus in the Mineral, Vegetable, and Animal Kingdoms, and the Biological Function of the Licithines," *ibid.*, XV (1893), 185–95. Maxwell's United States Department of Agriculture publications include: Walter Maxwell and Harvey W. Wiley, *Experiments with Sugar Beets in 1892*, USDA Division of Chemistry Bulletin, XXXVI (Washington, D.C., 1893); Walter Maxwell and Harvey W. Wiley, *Experiments with Sugar Beets in 1893*, USDA Division of Chemistry Bulletin, XXXIX (Washington, D.C., 1894).

12. for an outline of Schulze's career, see E. Winterstein, "Zur Erinnerung an Ernst Schulze," *Landwirtschaftlichen Versuchs-Stationen*, LXXVIII (1912), 303–20. To understand the relationship between the work of Schulze with that of Tollens, see E. Schulze and B. Tollens, "Untersuchungen uber das Holzgummi (Zylan) und die Pentosane als Bestanteil der inkrustierenden Substanzen der verholzten Pflanzenfaser," *Landwirtschaftlichen Versuchs-Stationen*, XL (1892), 367–78. See also E. Schulze, "Uber den Abbau und den Aufbau organischer Stickstoffverbindungen in den Pflanzen," *Landwirtschaftliche Jahrbucher*, XXXV (1906), 621–66; E. Schulze, E. Steiger, and W. Maxwell, "Untersuchungen uber die chemische Zusammensetzung einiger Leguminosensamen," *Landwirtschaftlichen Versuchs-Stationen*, XXXIX (1891), 269–326.

published over 180 papers during his long career, made important discoveries in the field of protein chemistry. He recognized that the protein transformations taking place during germination were fermentative processes. Further, Schulze and his students isolated and identified numerous complex nitrogen compounds in plants, including leucine, isoleucine, tyrosine, phenytalamine, proline, histidine, lysine, guanidine, glutamine, and arginine. Similarly, Maxwell's Louisiana studies centered around the dynamics of the transformation of organic molecules within living plants. Schulze's approach to plant biochemistry also proved to be an indirect influence on Beeson's work. It is apparent that Maxwell shared his former professor's ideas with his Louisiana colleague, since Beeson's studies on the physiological function of cane nodes was a reflection of Schulze's then-current research work on germination.

Maxwell's investigations in Louisiana were conducted in response to problems encountered during diffusion process trials at the station. He hypothesized that the presence of noncrystallizable organic bodies often interfered with the crystallization of diffusion juices. Diffusion, he thought, also extracted substances that were normally left in the bagasse during milling and left behind other substances usually extracted by pressure treatment.[13] In the past, nitrogen analyses conducted at the station assumed that all nitrogen was in the form of albuminoids. However, Maxwell performed a Kjeldahl nitrogen assay on mill and diffusion juices to determine total nitrogen, and then estimated amide nitrogen by the Stulzer method. Maxwell determined that while mill juices contained one-third albuminoids nitrogen and two-thirds nonalbuminoids (chiefly in the form of amides), diffusion juices contained one-fourth albuminoids and three-fourths nonalbuminoids. Maxwell's analytical work indicated that diffusion juices contained large quantities of amides. In the laboratory, samples of cane juice were first treated with a solution of subacetate of lead, then with hydrogen sulfide gas. The resulting liquor was treated with mercury nitrate, which precipitated the amide

13. Walter Maxwell, "Organic Solids Not Sugar in Cane Juices," "Sulphurous Acid, Acid Phosphate and Lime as Clarifying Agents," and "Fermentation of Cane Juices," in *The Chemistry of Sugar Cane and Its Products*, Louisiana Experiment Station Bulletin XXXVIII (Baton Rouge, 1895), 1371–1408.

bodies. Since albuminoid bodies were necessary for the formation of the coagulation blanket during the clarification process, it was now evident why hot water diffusion was often accompanied by crystallization and clarification problems.

Like his colleague Beeson, Maxwell related his findings to physiological interpretations of cane growth. He claimed that in mature seed, nitrogen was primarily in an albuminoid form; after the seed germinated, nitrogen was converted to the water soluble amide form. Thus, amide compounds, along with glucose, were easily transported to the various regions of the plant, where these substances furnished the raw materials for growth, while being reduced to albuminoid form once again.

Maxwell also studied various crude gums that were found in clarification tanks. These were generally regarded as being products of a process that was then denoted as viscous fermentation. He classified these substances into three groups: water soluble, soluble in dilute sulfuric acid, and soluble in cuprammonium. He found that these organic compounds were composed of modifications of starch and cellulose. Using phenylhydrazine to separate and to identify unknown carbohydrates, Maxwell found that these gums contained dextrose, levulose, and pentose sugars. Maxwell further classified gums into vegetable mucilages and true gums. When boiled with dilute acids, the hexosan bodies in mucilages converted to glucose sugars and cellulose, while pentosans in true gums yielded pentose sugars.

Maxwell's investigations opened new areas in sugar chemistry, particularly in his application of carbohydrate chemistry to the explanation of chemical reactions in the sugarhouse. His analytical studies of nitrogen compounds effectively marked the end of diffusion trials in Louisiana. While the diffusion process had proved itself as an effective method of obtaining high extraction yields, the addition of excess water resulted in increased fuel consumption. This dilution problem could have been surmounted, perhaps, by the use of efficient multiple-effect evaporation apparatuses. However, chemical process problems, particularly during crystallization, proved to be the death knell for the diffusion process. In an article in *Sugar Cane*, Maxwell remarked: "Diffusion yields the best results in proportion to the high grade of the cane. Where the juice has purity over ninety,

. . . the battery gets best results. Upon low grade cane, with juices of low purity, I believe diffusion is the worst possible practice."[14]

Thus, between 1890 and 1895 the research staff at the Louisiana Sugar Experiment Station concentrated on problems in sugar chemistry that were related not only to manufacturing difficulties, but also to important questions in plant physiology and nutrition. This was a new type of agricultural chemistry that was being practiced at the experiment station located at Audubon Park in New Orleans; it was quite different from the inorganic analyses that Stubbs had conducted between 1885 and 1889 at the station's previous location at Kenner. Unlike Stubbs, station chemists Beeson and Maxwell were apparently not satisfied with reporting the results of their work only in the experiment station's bulletins or in the *Louisiana Planter*. These chemists were eager not only to prove their work to the skeptical planter interest, but also to gain the recognition of the scientific community. Therefore, it is not surprising that Beeson communicated his findings also in the *Journal of the American Chemical Society*, nor that Maxwell published his results in the *Sugar Cane*.[15]

Routine, Research, and Retirement, 1896–1905

The nature of scientific activity conducted at the station changed markedly after 1896. Internal strife within the LSPA, the recession in the sugar industry, and the closing of the Audubon Sugar School resulted in decreased financial support for research. However, state assistance to the station increased between 1896 and 1905, enabling Stubbs to maintain a high level of scientific and technical activity in the laboratory. Yet, acceptance of state funding brought with it the obligation of increased regulatory analysis of soils, fertilizers, and pesticides, displacing original research. Beeson's and Maxwell's innovative studies examining the physiology of cane growth ended abruptly.

14. "Milling v. Diffusion," *Sugar Cane*, XXVII (1895), 428.

15. Beeson's experimental studies at the Louisiana Sugar Experiment Station were also published in the *Journal of the American Chemical Society*. See "Notes on the Estimation of Crude Fiber in Sugar Cane," *Journal of the American Chemical Society*, XVI (1894), 308–13; "A Simple and Convenient Extraction Apparatus for Food-Stuff Analysis," *ibid.*, XVII (1896), 744–45; "A Study of the Clarification of Sugar Cane Juice," *ibid.*, XIX (1897), 56–61.

With the passage of revised fertilizer legislation in 1898, new demands were placed on the sugar experiment station to conduct large numbers of analyses.[16] A manufacturer's guarantee of analysis and a state laboratory tag were required on each package or lot of fertilizer sold in the state. To comply with this legislation, a small corps of chemists was retained during both spring and fall to perform nitrogen, potassium, and potash analyses. Although only one chemist was responsible for this work in 1894, three chemists were assigned to regulatory analyses in 1900. After 1900 allocations for fertilizer analysis increased even more, from $2,350 in 1901 to $11,558 in 1904.

Assistant chemists were occupied with fertilizer analysis in spring and fall, but during the summer months their attention shifted to the chemical and physical analysis of soils. As director of experiment stations, Stubbs was also in charge of the State Geological Survey, and had directed geological investigations in several areas of the state prior to 1900.[17] After Stubbs entered into an agreement with Milton Whitney of the USDA Bureau of Soils to conduct a mapping survey of Louisiana in 1899, routine soil studies became an important activity of the sugar experiment station.[18] According to the contract, the USDA supplied trained field personnel and performed mechanical testing, while the Louisiana Experiment Station would be responsible for chemical analyses. The soils study ensured that the large group of assistant chemists would remain busy year round. In addition, a group of geologists joined chemists and agricultural scientists on the station payroll.

Stubbs found that the data obtained from these soil investigations were also valuable in understanding the process of plant growth. Mechanical properties of the soil proved to be important, since particle size could be correlated with moisture levels and the ability of plants to assimilate nutrients. In addition, bacterial content was found to be a critical property of nutrient uptake. This knowledge led Stubbs to reassess the design of implements and techniques used in preparing soils for cane cultivation. In 1891 James Mallon demonstrated

16. "Rules and Regulations and Suggestions Relative to the Purchase and Sale of Commercial Fertilizers," *LP*, XXI (1898), 194.

17. "Geological Survey of Louisiana," *LP*, XXIV (1900), 274–75.

18. Milton Whitney to the Secretary of Agriculture, June 23, 1899 (copy), in RG 164, Louisiana File, NA.

his rotary disc cultivator, which he claimed would give superior results in tilling and pulverizing the soil. Initially, Stubbs was skeptical of the value of such a device. However, the mechanical alteration of soil conditions improved yields and enhanced the quality
of the cane. Stubbs proposed that deep plows tended to decrease the
activity of beneficial soil bacteria, normally most active near the
surface. He felt that the success of Mallon's device was attributable
to its ability to maintain and even enhance bacterial activity within
the first few inches of the topsoil.[19]

While daily analyses occupied the majority of the chemists' time,
research studies investigating germination, plant composition, and
large-scale process reactions continued, but on a smaller scale. Non-
routine studies were often performed by chemists with bachelor's
degrees during the slack periods between the fertilizer and soil seasons. Research conducted after 1895 was generally imitative rather
than innovative. Well-honed analytical skills used in routine daily
analyses were then employed to perform careful assays on materials
not previously studied, and the resulting data were utilized to explain
physiological processes. For example, assistant chemists James E.
Halligan and H. P. Agee conducted a proximate analysis of sugarcane, comparing various constituents in the leaves, stalks, roots, and
seeds of the plant. Assistant chemist J. A. Hall studied the ash of
leaves, stalk, and roots, while Halligan and Joseph Verret examined
the varying proportions of fiber, sucrose, dextrose, levulose, ash,
acids, albuminoids, amides, and gums in cane as it matured.

After 1895 the Louisiana Sugar Experiment Station dropped from
the forefront of research. Sugar experiment stations in the West Indies and in Java became the primary source of scientific and technical methods that the station applied to problems facing Louisiana
sugar planters and manufacturers. Since the opening of the station,
Stubbs had attempted to cultivate numerous varieties of foreign cane
in hopes that he would find a sugar-rich plant suitable to the short
Louisiana growing season.

Previously it was thought that sugarcane, like many other plants,
reproduced only asexually from cuttings. However, in 1889 two British chemists employed at Dodd's Reformatory, a Barbados prison

19. "Preparation and Cultivation of Our Sugar Lands," *LP*, XVIII (1897), 170–72.

camp and experimental plantation, found twenty to twenty-five cane stalks that had "arrowed," or had developed a flowering seed-producing panicle.[20] The discovery of these fully matured plants with seeds stimulated a large-scale plant-breeding program in both Barbados and Java. This research resulted in the development of numerous new varieties of sugarcane, and seed samples were sent to several sugar experiment stations, including Audubon Park. Using the large Horticultural Hall at Audubon Park to grow these tropical varieties, in 1890 Stubbs tried to duplicate these plant-breeding experiments. Although these plants grew to enormous heights, they never matured. Later Stubbs received other hybrid seeds and two of these varieties, D 75 and D 95, proved to be richer in sugar content than the commonly planted La Pice cane. Excited by this finding, Stubbs lost no time in determining the manurial requirements for these plants and distributing cuttings to planters throughout Louisiana. He felt that the introduction of these hybrids was "the last hope in improving sugar cane. We have tried fertilizers of various kinds to increase tonnage . . . different varieties of cane . . . tile drainage . . . early fall [planting] . . . ; but we have . . . done better with these seedlings than with anything else."[21]

While developments from the West Indies had a significant impact on the Louisiana Sugar Experiment Station's activities, the emergence of a sugar research center in Java during the 1890s transformed the scientific basis of international sugar manufacturing. As a result, every aspect of the sugar business—manufacturing, pathology, cane breeding, engineering, entomology, and agriculture—was placed on the highest scientific level by a group of Dutch scientists employed in Java.[22] In particular, the work of botanist F. A. C. Went and chemist/engineer Prinsen Geerlings strongly influenced research efforts

20. William C. Stubbs, *Sugar Cane: A Treatise on the History, Botany and Agriculture of Sugar Cane,* 64.

21. "Seedling and Other Sugar Canes," *LP,* XXII (1899), 60–61.

22. Noel Deerr, *Cane Sugar* (London, 1911), vi; F. A. C. Went and H. C. Prinsen Geerlings, "On the Deterioration of the Sacchrose Content of Sugar Cane After Being Cut," *Sugar Cane,* XXVI (1894), 482–85; H. C. Prinsen Geerlings, "On the Carbonation of Cane Juice," *ibid.,* XXIX (1897), 229–33; H. C. Prinsen Geerlings, *On Cane Sugar and the Process of its Manufacture in Java* (Manchester, 1898). For an example of the influence of Javan developments on the Louisiana sugar industry, see *LP,* XXII (1899), 257.

in Louisiana. For example, Went's methods of determining the composition of sugarcane were used to analyze Louisiana cane in 1898. Also, Geerlings' manufacturing investigations, particularly his work on clarification, were pursued in Louisiana by the station's assistant director, R. E. Blouin. The research work of Charles A. Browne also reflected the strong influence of the Dutch upon research conducted at the Louisiana Sugar Experiment Station during the early 1900s.

Hired by Stubbs in 1902, Browne, formerly of the Pennsylvania Experiment Station and a student of Tollens' at Göttingen, came to Louisiana to "make original investigations along the line of carbohydrates and other bodies which occur in small quantities in sugar cane juice or its products."[23] With the assistance of several chemists, Browne followed Went's example in conducting careful determinations of water, ash, fat, wax, nitrogenous compounds, fiber, and sugars in leaves, stalks, roots, and seed.[24] Furthermore, in an attempt to quantify the amount of inorganic nutrients that cane extracted from the soil, he assayed plant ashes for K_2O, Na_2O, CaO, MgO, Fe_2O_3, Al_2O_3, SiO_2, P_2O_5, SO_3, CO_2, Cl, and C.

After several months at the station, Browne's research interests shifted to determining the role of enzymes in plants. As in the past, manufacturing and cultivation problems provided the practical justification for a scientific study. In this case, manufacturing losses from fermentation seemed to be linked with the process of ripening and subsequent deterioration of cane. He used the 1893 investigation of M. Raciborski in Java as the starting point for his research. By the use of a solution of guaiacum blue, Raciborski had detected the presence of oxidases in longitudinally cut sections of cane.[25] Drawing from Went's explanation of the role of enzymes in plants, Browne asserted that enzymes in sugarcane converted starches and sugars to reducing sugars, which were then transported to points of growth and, in turn, employed in the creation of new tissues. In an experiment similar to that conducted by Went and Geerlings, Browne de-

23. W. C. Stubbs to C. A. Browne, October 13, 1902, in Browne Papers, Box 15, LC.

24. C. A. Browne and R. E. Blouin, *The Chemistry of the Sugar Cane and Its Products in Louisiana,* Louisiana Experiment Station Bulletin XCI (Baton Rouge, 1906), 15.

25. M. Raciborski, "On Some Constituents of the Sugar Cane," *Sugar Cane,* XXX (1898), 474–78.

termined the rate of conversion of dextrose and levulose (reducing sugars) to sucrose in maturing cane stalks.

While Browne was given the opportunity to conduct chemical research that was of interest to the scientific community, he was also constantly directed to apply his scientific expertise to the solution of practical problems.[26] In that vein, Browne examined the relationship between various types of bacterial contamination and undesired reaction products in the sugarhouse.[27] In addition, his skills in organic analysis led to a comprehensive understanding of the hydrolytic products of bagasse. It was hoped that this knowledge would enable sugar interests to develop an economical method for utilizing this waste product in paper manufacturing.

Browne had far broader responsibilities than merely conducting research experiments at the station. Like Maxwell, Browne studied various clarification processes, including those using lime, sulfur dioxide and lime, superheat, and electrical currents. In addition, he was frequently called upon to speak at farmers' institutes and LSPA meetings. Finally, he was also charged with the preparation of exhibits for the 1904 Louisiana Purchase Exposition at St. Louis.[28] This was not a new role for the staff of the sugar experiment station. Previously, Stubbs had organized the Louisiana agricultural exhibit at the Pan-American Exposition held in Buffalo, New York, in 1901. Thus, for a number of the station staff, public relations was an important part of their responsibilities, in addition to conducting scientific research. Portraying Louisiana as a prosperous and progressive state, the exhibits allowed exposition visitors to view large models of the latest designs of sugar-manufacturing equipment and a wide variety of the agricultural products found in the state. It was hoped that such efforts would attract settlers to Louisiana.

Besides conducting agricultural research, the station had other ob-

26. C. A. Browne to A. W. Dykers, August 17, 1940, in Browne Papers, Box 4, LC.

27. C. A. Browne, "The Effects of Fermentation Upon the Composition of Cider and Vinegar," *Journal of the American Chemical Society,* XXV (1903), 16–33; C. A. Browne, "The Analysis of Sugar Mixtures," *ibid.,* XXVIII (1906), 439–52; C. A. Browne, "The Fermentation of Sugar-Cane Products," *ibid.,* XXVIII (1906), 453–69.

28. Louisiana Board of Commissioners to the Louisiana Purchase Exposition, *Louisiana at Louisiana Purchase Exposition* (New Orleans, 1904).

ligations to the state: regulatory analyses, public relations, and education. However, Stubbs's reluctance to include teaching in the station's functions eventually led to his initiating a bitter campaign to separate the experiment station from the state university. He felt that teaching responsibilities diffused the staff's research efforts. "The station should bend its energies on research work and the scientific facts obtained should be furnished to the professors in the Agricultural College to be included in their teachings. No station worker should be connected with teaching in any way. The work of the station and the university should be as distinct as possible."[29] Stubbs's views concerning the teaching obligations of station personnel were contested by Thomas Boyd, president of Louisiana State University. Perhaps it was the disagreement concerning the role of the experiment station that caused Stubbs to submit his resignation as experiment station director in 1902. Although Stubbs was apparently content to remain at Audubon Park as assistant director, Boyd persuaded him to continue as director.[30] This decision may have been influenced by Boyd's partial acquiescence to Stubbs's demands to separate the station from the university. In 1903 Stubbs wrote to A. C. True that "we are concentrating our work more along station lines, dissociating, as far as we possibly can, our station workers from teaching. After January next, Prof. Morgan will be connected entirely with the station, and will no longer have any connection with the college. . . . There are only two professors now connected with the station and the university."[31]

While Stubbs was able to negotiate a satisfactory working arrangement with Boyd, apparently he could not with Governor Newton C. Blanchard. Stubbs publicly announced that in order to regain his health he wished to retire. However, it was privately rumored that the station director and the governor had had a disagreement and that Stubbs had promised to resign at the conclusion of the St. Louis exposition.[32] True to his word, Stubbs retired in November, 1904.

29. Undated letter, Folder 17, in Stubbs Papers, SHC.

30. Thomas D. Boyd to Alfred C. True, March 8, 1902, in RG 164, Louisiana File, NA.

31. Stubbs to True, December 2, 1908, in RG 164, Louisiana File, NA.

32. "Is Dr. Stubbs Getting Bad?" [1905?] (newspaper clipping in Stubbs Papers, Folder 20, SHC).

Stubbs proved to be a crucial figure in the development of the Louisiana sugar industry. As a member of the executive committee of the LSPA and director of the sugar experiment station, he clearly perceived pressing agricultural and processing problems in Louisiana and used these problems to guide station investigations. Stubbs's own research interest in plant growth was sufficiently broad to incorporate knowledge gained from both practical manufacturing studies and analytical chemical research.

Between 1885 and 1905 the agricultural chemistry practiced at the Louisiana Sugar Experiment Station changed markedly. Stubbs's approach had its basis in the work of mid-nineteenth-century chemists. His view of agricultural chemistry was technically and symbolically rooted in the soil and focused on the interests of the local planter class. After 1890 a new generation of scientists found employment at the Louisiana Sugar Experiment Station. These European-educated chemists emphasized organic structural chemistry and plant biochemistry in their research work. Unlike Stubbs, who was closely tied to the culture and institutions of the South, these new scientists sought international recognition for their work and often found employment in other cane-producing areas. Between 1885 and 1905 the research program at the Audubon Park station not only served local interests, but also made contributions that were valued by the international scientific community.

10

Professionalization and Engineering Education 1890–1905

The formation of the Louisiana Sugar Planters' Association and subsequent institutional developments between 1877 and 1900 were a reflection of developments occurring in a dynamic, expanding industry. During this period many aspects of the sugar industry were in flux, as individuals and groups gradually worked out stable relationships and practices to suit the conditions of a new economic environment. The process of reorganization and the establishment of a new order were most evident among professional sugar chemists in Louisiana. Whether employed in the sugarhouse, experiment station, or university, the Louisiana sugar chemist faced a number of professional, status-related problems. To begin with, unlike the practical engineer who had provided valuable services to the industry since the 1830s, the chemist now found himself in a new industrial setting, frequently assuming responsibilities that were once associated with the sugar boiler. However, many planters still considered the theoretical training of educated station chemists as generally insufficient to deal with the realities of production. Further, the LSPA did not actively encourage the exchange of chemical information or promote the chemical profession, since its meetings normally focused on engineering problems, rather than topics in sugar chemistry.

Professionalization and the Louisiana Sugar Chemists' Association

In an effort to enhance their professional status, factory chemists, normally in charge of process control, led an early movement to organize a professional association for sugar chemists in Louisiana. In 1889 Lezin Becnel, a chemist employed at the McCall brothers' plantation in Ascension Parish, took the initiative to establish a local chemists' society.[1] His actions were primarily in response to the perceived widespread opposition to sugar chemists by both skilled and unskilled laborers. Becnel hoped that an association would clearly define the role of the chemist in the sugar industry, as well as assist in uniting chemists and planter/employers for their mutual benefit. Becnel asserted that the factory chemist was a misunderstood and frequently maligned professional whom plant workers often called a "crank" working in a "drugstore."[2]

Yet Louisiana chemists were not as annoyed by name-calling as by overt resistance on the part of skilled workers to their presence in the sugarhouse. The first paper read to the Louisiana Sugar Chemists' Association in 1889 dealt with this conflict. H. L. Youngs, another plant chemist, complained that process records were often falsified by workers who feared that the chemist was evaluating their performance and displacing them from their traditional duties. Youngs stated that the skilled laborer in the sugarhouse

> seems to have an innate antipathy for the chemist, and in every way possible tries to deride his efforts or detract from the merits of his labor; and all of this from a mistaken idea of the cause of the chemists' presence in the sugar house. He considers him a sort of detective engaged purposely to watch the sugar maker, and record his every action with a view to his condemnation. If the chemist orders more time in clarification, the sugar maker complains that he can not make white sugar or that the molasses will be too black—anything for an excuse to disagree with the chemist.[3]

1. "Louisiana Chemists Organizing," *LP*, II (1889), 270. To understand Becnel's work, see Lezin A. Becnel, *Report on the Results of Belle Alliance, Evan Hall and Souvenir Sugar Houses, for the Crop of 1888* (New Orleans, 1889).

2. L. A. Becnel, "General Plantation and Sugar House Statistics, the Manner of Keeping Same, and the Necessary Chemical Control," *LP*, II (1889), 286–87.

3. "Louisiana Sugar Chemists' Association," *LP*, IV (1890), 263–64.

While factory chemists were promoting and clarifying their position in the *Louisiana Planter and Sugar Manufacturer*, experiment station chemists recognized that the theoretically guided scientist suffered from a lack of prestige within planter circles. Louisiana planters emphasized that their business needed the contributions of practical chemists. According to this view, the successful chemist possessed "a large stock of practical knowledge, and of experience in actual sugar house work. . . . To employ a professor of chemistry unpossessed of this experience would be to lose a season."[4] Thus, both industrial and station chemists had sufficient reasons to form an association. The undetermined role of the chemist, and the low professional regard in which he was held, united these two groups of chemists possessing somewhat different educational backgrounds and professional objectives. While all of the experiment station chemists had earned at least bachelor's degrees, Lezin Becnel, the leader of the sugarhouse chemists, had no formal training in chemistry. Becnel had begun his career as a foundry apprentice, and had learned sugar chemistry through practical experience. Experiment station chemists viewed their role in the development of the Louisiana sugar industry in terms of perfecting and approving standard analytical methods. In contrast, many sugarhouse chemists, like Becnel, felt that their efforts should be focused on developing standard record-keeping techniques to effect the scientific control of the sugar manufacturing process.[5] This emphasis upon "scientific bookkeeping" had its origins in the German beet-sugar industry and was transplanted to Louisiana during the USDA experiments of the 1880s.

On June 15, 1889, the Louisiana Sugar Chemists' Association held its first meeting at the Louisiana Sugar Experiment Station.[6] A number of the station's staff, including W. C. Stubbs, were in attendance, and B. B. Ross of the Baton Rouge station was elected the group's first president. The association's objectives included the dissemination of chemical knowledge, the discussion and evaluation of analytical methods, and the adoption of a standard system of chemical control statistics for the factory. Membership to this organization was limited to chemists who were either employed in the analyses of agri-

---

4. "Notes and Comments," *LP*, III (1889), 370.
5. For an example of chemical control, see Hubert Edson, *Calumet Sugarhouse Results, Campaign of 1889–90* (Louisville, Ky., 1890).
6. "The Sugar Chemists," *LP*, II (1889), 283.

cultural products or in the control of sugarhouse processes. The organization was led by its officers and two permanent three-member committees. The committee on statistics was selected from practical sugarhouse chemists, while the committee on analysis was comprised of station chemists.

The first priority of the newly formed chemists' association was the development of close ties with the LSPA. Indeed, Henry Clay Warmoth was not only in attendance at the first session, but he also addressed the newly organized scientists. Later in the evening after the conclusion of the first meeting, the chemists made it a point to attend the LSPA gathering, where they informed planters of their new organization. That same evening the LSPA passed a resolution initiated by Henry McCall which stated that "we heartily approve of the formation of this new organization, and that we will urge the adoption of standards of all the methods of keeping plantation and sugar house data recommended by the Chemical Association."[7]

During the summer of 1889, led by association president B. B. Ross and Maurice Bird of the sugar experiment station, the committee on analysis compiled a list of the most satisfactory analytical procedures for the determination of sugar. The fact that they published their recommended procedures and research data in a widely circulated pamphlet that fall indicates that these station chemists were attempting to establish themselves as experts within the Louisiana chemical community. The booklet, entitled *Report of the Committee on Methods of Sugar Analyses of the Louisiana Sugar Chemists' Association*, provided detailed descriptions of the methods and apparatuses necessary for the determination of sucrose, specific gravity, density, total solids, glucose, ash, and fiber. This step-by-step outline was undoubtedly written for the benefit of analysts who were "just entering [the profession] . . . , or . . . [who had] only a limited experience in the laboratory of the sugar house."[8]

7. "Regular Meeting of the Louisiana Sugar Planters' Association," *LP,* II (1889), 283.

8. *Report of the Committee on Methods of Sugar Analyses of the Louisiana Sugar Chemists' Association* (Baton Rouge [1889]), 201. This pamphlet was reprinted in the *Journal of Analytical Chemistry,* IV (1890), 1–19. The proposed revision of the constant used in Clerget's method caused a controversy among analytical chemists. See "Changes of Methods of Analyses by the Association of Official Agricultural Chemists," *Journal of Analytical Chemistry,* VI (1892), 259–62; "Concerning the Constant

The committee had attempted to establish itself as an authority by approving the numerous chemical procedures used in the chemical control of the factory. In addition, the committee of station chemists conducted an exhaustive study on Clerget's method for the determination of sucrose in the presence of glucose. Precise polariscopic studies were employed to ascertain a more accurate value for the constant in Clerget's mathematical relationship correlating sucrose concentration with optical rotation. Even a small error in this constant could affect the profits of large-scale sugar processors, since the market price for molasses was determined by its sucrose content. Following Clerget's procedure, the analyst first treated a sample containing a mixture of sucrose and reducing sugars with hydrochloric acid, which subsequently caused the sucrose to react with water and form both glucose and fructose. In 1849 Clerget had found that the amount of change in the rotation of polarized light was directly proportional to the initial concentration of sucrose, and that one normal weight of this substance at 0°C caused a 144-degree change in the polariscopic reading. Their extreme care with reagents and experimental conditions, as well as their treatment of data and the fact that they also published their results in the *Journal of Analytical Chemistry*, suggests that Ross and Bird were also hoping to earn the respect of the American scientific community. Their experiments, which utilized a double-compensating polariscope, were conducted at a temperature of 31°F in a room at the New Orleans Cold Storage Company. To verify optical analyses, results were compared with data obtained from gravimetric determinations using Fehling's solution.

Concurrently, the association's statistics committee began to devise a standard form for reporting chemical process information. It was hoped that this form would eventually become the basis for

---

to be Used in Clerget's Inversion Process," *ibid.*, VI (1892), 432–35, 519–24, 633–36. Clerget's procedure, first proposed in 1849, enables the analyst to determine the percentage of sucrose in a solution containing other optically active substances. Cane sugar (sucrose) is hydrolyzed by either acid or the enzyme invertase to a 1:1 mixture of glucose and fructose, which rotates polarized light in the opposite direction. The amount of change in rotation is proportional to the concentration of sucrose. The exact amount of change was the subject of careful research by the Louisiana sugar chemists. For additional information on Clerget's method, see C. A. Browne and F. W. Zerban, *Physical and Chemical Methods of Sugar Analysis* (New York, 1941), 402–405.

comparing the performance of various Louisiana sugarhouses, and for indicating manufacturing inefficiencies. Becnel asserted that in the future "our records should also include the cost of every item, from the cutting of the cane to the cost of laying down one pound of sugar on our plantation landings, so as to . . . reduce the actual cost of manufacture."[9] The introduction of chemical control in the Louisiana sugar industry ultimately revolutionized business practices. For example, once John N. Pharr employed a chemist at his Glenwild factory, annual reports became more detailed and manufacturing costs were calculated. Pharr's operation divided expenses into several main categories, including labor, board, laboratory, illuminants, lubricants, lime, acids, filter cloths, wood, and coal. No doubt this data contributed to a reduction in Pharr's manufacturing costs from $111.00 per ton in 1894 to $93.65 in 1900.[10] A similar type of worksheet accounting system was employed at the Gay family plantation, where unit costs were determined for material handling, crushing, clarifying, evaporating, and other operations.

The Louisiana Sugar Chemists' Association was essentially comprised of two groups of chemists, each pursuing activities that furthered their own interests. The organization met in the fall and spring, and discussions usually consisted of reports from various members on works in progress. In 1892, however, events outside the association led to a definite shift in the organization's objectives. A series of attacks by planters, engineers, and sugar boilers questioned the effectiveness of chemists in general, and the association in particular. One planter asserted, "I will state that so far as I know, no one single method or improvements of an old one, has been originated by the Association."[11] An anonymous critic commented:

> What have chemists done for the manufacture of sugar from cane these last 30 years, other than neutralizing the acid in the juice of cane and making the proper defecation by the use of lime? As far as I can learn, all

9. Becnel, "General Plantation and Sugar House Statistics," 287.

10. "Report of the Committee on Blanks of the Louisiana Sugar Chemists' Association, September 24, 1892," *LP*, IX (1892), 250. See "Laboratory Notebook 1891–1895" and "Comparison of Seven Years Nine Roller Mill Work" (MSS in John N. Pharr and Family Papers, Special Collections, Hill Memorial Library, LSU); "Plantation Report Book" (MS in Edward J. Gay and Family Papers, Special Collections, Hill Memorial Library, LSU), 108–10.

11. De Nos, "Sugar House Laboratories," *LP*, IX (1892), 38.

the improvements that have been made in the industry have been mechanical. The first have been in securing good extractions; secondly, in improved multiple effects for evaporation and in improved vacuum pans and centrifugals. I am not necessarily prejudiced against chemists, but they seem to claim the world just now, and I inquire to learn what they have really done. So far as I know they have done but little.[12]

A sugar boiler also vented his frustrations.

If the chemist would use plain, commercial English, and quit compounding words that Webster never heard of, he would be of more value to the common sugar-house man. . . .

Let them answer these questions and quit condemning the ignorant sugar boiler, the careless engineer and the willful, wasteful sugar house hand in general.

According to the chemist there is not a man employed in Louisiana sugar houses capable of cleaning the trash from under the cane shed without the chemist being there to see that he does not waste something. It seems strange to me that so many sugar boilers should ever find employment considering their ignorance.

All the improvements I can see in the sugar industry are the five and six roller mills, the double effect, the filter press, the bagasse burner and the improved centrifugals, all of which are inventions of our engineers. The chemist has told us how much sugar is in the cane, but he has never invented anything to get it out or even told us how to work frozen cane, or what to do with juice that no reasonable amount of lime will clean.[13]

In defense, J. T. Crawley, a Harvard-educated chemist at the Louisiana Sugar Experiment Station, responded to the criticisms by claiming that many sugarhouse chemists in Louisiana were really only part-time employees who remained in the state only during the three-month harvest season.[14] Therefore these men were not in a position to make research contributions. Crawley remarked that Louisiana sugar chemists were so burdened with routine work that necessary investigations to improve manufacturing processes and field techniques were in reality impossible. Yet, the association took deliberate steps to stimulate research and answer their critics' charges concerning their profession's inability to innovate. In 1892, the asso-

12. "What Have Our Chemists Done?," *LP*, 8 (1892), 388.
13. "The Sugar Chemists Again," *LP*, VIII (1892), 445.
14. "Sugar Chemistry," *LP*, VIII (1892), 237.

ciation created a library fund, since first-rate research required access to current knowledge. A number of important journals were ordered by the membership, including: *Sugar Cane, La Sucrerie Indigène et Coloniale, Bulletin de l'Association Belge des Chimistes, Scheibler's Neue Zeitschrift für Rübenzucker Industrie, Zeitschrift für Analytische Chemie,* and *L'Alcool et le Sucre.*

With the association's acquisition of this literature, more of its members began to apply European analytical and agricultural chemistry techniques in their research projects.[15] Hubert Edson, former USDA chemist and nephew of Harvey Wiley, based his studies of fertilizers and sucrose inversion on investigations that were originally published in *Bulletin de l'Association des Chimistes, Compte Rendus, Annales Agronomie,* and *Chimie Agricole.* Remarkably, industrial chemists like Clinton Townshend now contributed well-organized, specific research papers, rather than the rambling discourses given previously. Station scientists also improved the quality of their original research. For example, B. B. Ross began to develop an indirect method for the determination of inverted sugars that employed electrolytic copper methods. This technique had been suggested in work appearing in the *Chemiker Zeitung* and the *Zeitung für Rübenzucker Industrie.* Clearly, the topics discussed at the sugar chemists' meetings were no longer comprehensible to the average planter, but rather were directed at a local scientific audience.

Concurrent with the movement to stimulate research within the membership, the chemists' association began to take steps to identify itself within a larger scientific community. As early as 1891 the organization had expressed interest in joining a national group of chemists centered in Washington and led by Albert Peale of the United States Geological Survey. By 1893 most of the members had already joined the American Chemical Society (ACS), based in New York City, and in late 1893 the association officially took steps to join that body. In 1894 the Louisiana Sugar Chemists' Associa-

15. Hubert Edson, "The Effect of Glucose on Sucrose Crystallization," *LP,* X (1893), 100–101; Hubert Edson, "Cane Fertilization," *LP,* XII (1894), 105–107; Clinton Townsend, "Glucose in Its Relations to Sugar Manufacture," *LP,* X (1893), 100; Clinton Townsend, "Double Polarization," *LP,* X (1893), 265–66; B. B. Ross, "The Electrolytic Estimation of Copper as Applied to Invert Sugar Determinations," *LP,* X (1893), 139; J. T. Crawley, "Effects of Sulphur on Clarification," *LP,* X (1893), 380–81.

tion became a part of the newly formed New Orleans section of the ACS. This local branch was short-lived, however, operating only until 1898.[16]

While chemists contributed to the revolutionary changes taking place in the Louisiana sugar industry in the late nineteenth century, engineering innovations were at the center of this industrial transformation. During the last two decades of the nineteenth century, the scale, methods, machinery, and structure of the Louisiana sugar industry changed dramatically. In 1880 there were approximately one thousand sugarhouses, each producing an average of 110 long tons of sugar annually. By 1900 less than three hundred factories made up Louisiana's sugar industry, but yearly production averaged over 980 long tons for each sugarhouse. While animal-powered mills and open kettles were common in the 1880s, relatively few were to be seen at the turn of the century. This new industrial order was accompanied by an increased reliance on chemical and engineering knowledge, improved business and accounting practices, and the gradual separation of cane-growing and manufacturing activities.

During the late 1880s a number of Louisiana planters invested in new machinery and process apparatuses. Confident of continued tariff protection and cognizant of the worldwide trend of declining sugar prices, such planters as Lewis Clarke and H. C. Warmoth installed large-capacity diffusion cells and triple-effect evaporators in their plantation sugarhouses. John N. Pharr's actions were typical of the changing attitudes toward investment in process machinery. Pharr, a St. Mary Parish planter, had employed the same inefficient three-roller mill between 1878 and 1888. However, in 1889 he decided to purchase nearly $100,000 worth of new equipment, including a diffusion battery, an eight-foot-diameter vacuum pan, a Lille triple-effect evaporator, two roller mills, and a narrow gauge plantation railroad. Pharr's new sugarhouse was illuminated with electricity, and its processes were monitored by a chemist who prepared daily production records and an annual report.[17]

16. Between 1896 and 1898 the inside cover of the *Journal of the American Chemical Society* listed a local New Orleans chapter. In 1900 the New Orleans section was no longer included, and in 1906 a second New Orleans chemists' association affiliated with the American Chemical Society was established.

17. Irving P. Foote, "A Louisiana Sugar Plantation and the Sugar Industry: A Type Study" (M.A. Thesis, George Peabody College for Teachers [1922?]).

This rapid pace of industrial development was catalyzed by the passage of federal bounty legislation in 1890. The Republican-sponsored McKinley Act, which was to be in effect for fifteen years, stipulated that sugar would be placed on the free list, but it also provided for a drawback of $0.02 per pound on vacuum pan sugars and a subsidy of $0.0125 per pound on open-kettle products.[18] Although once reluctant to invest in machinery, many sugar planters "went wild, and went into lavish and extraordinary expenditures."[19] Efficient manufacturers with advanced-design apparatuses could produce sugar for about $0.016 per pound, and thus could profit handsomely.[20]

A large surplus in the federal treasury provided the funds for the sugar bounty payments. However, in 1892 the Democratic Cleveland administration came to power. Subsequently, steps were taken to abolish the bounty provisions contained in the McKinley Act, which critics had cited as legislation benefiting only a few at the expense of the majority of American citizens. Over the protestations of the Louisiana congressional delegation, the bounty was repealed in 1894. Yet, in the four years of its existence, over thirty million dollars had been paid to Louisiana planters, providing the stimulus for the creation of a large-scale industry whose modern plants were both scientifically controlled and technically designed.

The LSPA reacted to this boom in machinery investment by serving as a forum for the discussion of the merits of various newly designed apparatuses. During the early 1890s more than 50 percent of LSPA meetings were devoted to foundrymen's discussions of topics like "The Best Boilers for Plantation Use," "The Horsepower of Steam Boilers," "The Economic Results of Pipe Covering," "The Relative Merits of Multiple Effects," "The Proper Construction of Hot Rooms for Sugar Houses," "The Clarification of Cane Juices," "The Burning of Bagasse," "Steam Economy and the Corliss Engine," "The Proper Method of Sulphuring Cane Juices . . . ," and "Best Plans and Specifications for a Model Sugar House." In addition to holding technical discussions, the association concurrently devel-

18. "The Sugar Bounty," *LP*, V (1890), 227–28; "What We Must Do to Realize the Bounty on Sugar," *LP*, V (1890), 347–48.

19. Henry McCall, "History of Evan Hall Plantation," 1924 (Typescript in Special Collections, Hill Memorial Library, LSU), 19.

20. "Annual Report, 1894," and "Record Book, 1891–1928," both in John N. Pharr and Family Papers, LSU.

oped a new type of scientific and engineering education in Louisiana. Until the decade of the 1880s, the overseer, sugarhouse engineer, and sugar maker were the skilled workers associated with the sugar industry. These artisans employed knowledge that was usually acquired on the job, normally through apprenticeship training. Information on such operations as boiling the syrup to crystallization or clarifying the expressed juice was informally passed from one generation to another and could not be found in technical treatises. This method of educating sugarhouse personnel was particularly disturbing to LSPA elite like John Dymond, who felt that this way of learning put planters at a disadvantage to the European competition. Dymond remarked, "If each one studies for himself and imparts no knowledge to others, and learns nothing from others, we shall soon be left behind in the race."[21] With the rapid modernization of the sugar industry, Louisiana planters and manufacturers called for the training of a new kind of professional to manage the sugarhouse.

The Audubon Sugar School, 1891–1896

The Audubon Sugar School, with its emphasis on practice, was established in 1891 in response to the perceived lack of scientific expertise in the Louisiana sugar industry. Despite LSPA efforts, both Tulane University and Louisiana State University had failed in developing undergraduate technical programs to meet the needs of the sugar interest during the 1880s.[22] Thus, in 1888, John Dymond, Plaquemines Parish planter and editor of the *Louisiana Planter and Sugar Manufacturer*, called for the establishment of a private technical school at the sugar experiment station.[23] Dymond planned to employ the same fund-raising methods as had been used so successfully in the establishment of the Kenner experiment station in 1885. He argued that if one hundred planters each contributed one hundred dollars, W. C. Stubbs, director of the station, would have ample funds at his disposal to hire the necessary faculty.

Stubbs had already initiated an informal training program at the station, but he was far from pleased with its results. He was particu-

21. "Sugar Schools," *LP*, I (1888), 253.

22. Louisiana Sugar Planters' Association, *Regular Monthly Meeting, September 1887* (N.p., n.d.), 24.

23. "Sugar Schools," *LP*, I (1888), 253.

larly annoyed that many volunteer workers spent only a few months familiarizing themselves with the routine analytical procedures and methods of chemical control before they left for employment opportunities. He felt that these trainees never attained the required level of expertise to be effective sugarhouse managers. In addition, budget limitations precluded Stubbs from offering all qualified individuals an opportunity to receive practical instruction. It was hoped that a well-funded school would enable planters' sons to study at the station, "to learn, with hammer and chisel, polariscope and crucible, actually in hand, how to benefit themselves and their favorite industry, and perhaps to preserve the homes of their fathers from encroaching debts, the legacy of the old regime."[24]

Although Dymond's efforts in 1888 proved ineffectual, he continued to campaign for the founding of a sugar school in Louisiana. He maintained that the rise of the science-based and economically efficient European beet-sugar industry posed a serious competitive threat to the rule-of-thumb Louisiana planter and manufacturer. According to Dymond, the future success of the Louisiana sugar industry depended on its ability not only to adopt European methods and apparatuses, but also to create institutions to facilitate the diffusion of this knowledge. He remarked: "If our German cousins are doing all this [scientific education] in their old-fashioned prosaic sort of country, what should we be doing here in progressive America, the home of invention and the cradle of energy? We should have sugar schools, and we should have them at once. And let us have one forthwith here in Louisiana."[25] Dymond viewed the school's graduates as the cadre of an industrial revolution—one in which the tradition-bound Louisiana sugar industry would be rapidly transformed into a scientifically controlled business with the large central factory as its focal point.

Dymond was not the only American urging the formation of a sugar school. Harvey Wiley was promoting the establishment of a similar institution, not to be located in Louisiana, however, but at the USDA's new laboratory in Washington. Wiley enlisted the support of Senator E. H. Funston of Kansas, who proposed an amendment to the 1890 agricultural appropriation bill that would pro-

24. *Ibid.*
25. "Sugar Schools," *LP,* II (1889), 85.

vide funds for an institution to train young men in the cultivation and manufacture of sugar from cane, beets, and sorghum. Wiley proposed that upon the successful completion of an entrance exam, students would spend from one to three years in acquiring both classroom knowledge and practical experience. However, Funston's legislative efforts were defeated.[26]

In 1890 Dymond's educational ideas came to fruition as a consequence of a favorable economic climate and the possibility of obtaining federal funds. Planters appeared to be confident that the bounty regulations contained in the McKinley tariff would ensure their prosperity for at least fifteen years. In addition, the passage of the 1890 Morrill Act raised the possibility that the federal government would contribute funds for the establishment of a local sugar school. The Morrill Act stipulated that a portion of the proceeds from land sales would be used to endow and support state agricultural and mechanical colleges. Louisiana planters had reason to be optimistic, since the legislation clearly stated that the grant be applied "only to instruction in agriculture, the mechanic arts, the English language, and the various branches of mathematical, physical, natural and economic science, with special reference to their application [to] . . . industries."[27]

Further encouraged, Dymond outlined his organizational plans for the sugar school at the LSPA's monthly meeting in October of 1890. Dymond stated that important American industrial interests—railroad, steel, and mining—had successfully tapped a vast reservoir of technical talent produced by institutions like Rensselaer Polytechnic Institute, Stevens Institute, Columbia College School of Mines, Yale, and Harvard.[28] He suggested that Louisiana initiate the cooperation seen among universities and manufacturers in the Northeast by founding an institution specifically to serve the sugar interests. This institution would supply sugar manufacturers with trained experts to ensure the industry's survival in the competitive world market. He proposed that the curriculum be focused on three critical areas—agriculture, chemistry, and mechanical engineering—and that the two-year program stress both theory and practice. The pro-

26. Wiley to E. H. Funston, January 14, 1890, in RG 97, NA.
27. "The Agricultural College Bill," *LP,* V (1890) 289.
28. "A Sugar School," *LP,* V (1890), 280; Louisiana Sugar Planters' Association, "Minutes of Monthly Meetings, 1877–1891," October 9, 1890, p. 297.

gram's graduates would be capable of uniting these disparate bodies of knowledge in their daily activities, and thus exercise "one-man power and with full control, bending every effort to reach excellence and economy at the same time."[29] They would apply their scientific training not only to perfect the traditional methods of the sugar boiler, whose chief function was to produce high quality sugar, but also to improve the empirical techniques of the sugarhouse engineer, who was responsible for reducing production costs. Dymond claimed that these artisans were often working against each other, and that the union of their skills would centralize authority with one man, the sugarhouse superintendent.

The LSPA membership reacted favorably to Dymond's message, and the association formed an eight-man committee to negotiate with the Louisiana State University Board of Supervisors concerning the use of Morrill funds for the creation of a sugar school.[30] However, typically short of funds and involved in political turmoil, the university was apparently reluctant to share the newly acquired federal monies with the sugar planters. Thus, as in the case of the experiment station, the planters shouldered the initial financial responsibilities for the formation of the school. In June, 1891, the Executive Committee of the Louisiana Scientific Agricultural Association, led by its chairman, John Dymond, organized the Audubon Sugar School at the sugar experiment station in New Orleans. In addition to the proposed two-year course, which was scheduled to begin in the fall of 1891, provisions were made to accept special students desiring less comprehensive training.[31]

The influence of the German sugar-beet industry on the development of the Louisiana school was significant. The curriculum and the background of the students at the early Audubon school were similar to that of the two-year school located at a sugar experiment station in Brunswick, Germany. This similarity is not surprising, since the Louisiana sugar interests were well acquainted with the German school; its annual report was published yearly in the *Louisiana Planter and Sugar Manufacturer*.[32] Also, many planters had

29. "Sugar House Superintendents," *LP*, V (1890), 487.
30. "The Sugar School and Other Matters," *LP*, V (1890), 298–99.
31. "A Sugar School," *LP*, VI (1891), 457.
32. "Sugar School at Brunswick, Germany," *LP*, II (1889), 39; "The Sugar School at Brunswick, Germany," *LP*, VI (1891), 111; "Sugar Schools," *LP*, X (1893), 114; "Sugar

been favorably impressed with the skills of Frederick Hinze, a Brunswick sugar school graduate who had been employed by Warmoth during the USDA experiments of the 1880s.[33] The majority of the Brunswick students were in their late twenties and already had previous experience in the sugar trade. Initially founded by a group of local manufacturers, the Brunswick school later received government assistance. Students studied four basic areas—analytical chemistry, sugar technology, agriculture, and commercial science.[34]

However, the Audubon Sugar School was not an exact copy of its German counterpart, for Stubbs's own views on education also helped shape the structure of the institution's curriculum. In reconciling his bitter memories of the Civil War, he placed the blame for the South's defeat on its educational system. In Stubbs's opinion, the classical curriculum of antebellum southern colleges had trained an elite group of jurists, scholars, and men of culture, but it had failed to create the necessary class of mechanics, farmers, and businessmen. Stubbs asserted that this lack of expertise resulted in the mismanagement of southern resources between 1861 and 1865 and eventually led to the South's surrender at Appomattox. He was convinced "that to meet the wants of the New South, our institutions of higher learning must no longer alienate . . . the industrial interests of our country."[35]

In addition to his personal opinions regarding the Civil War, Stubbs's ideas also reflected widespread views concerning the type of education that should be taught at the agricultural and mechanical colleges created by the 1862 Morrill Act. Between 1840 and 1860 a number of educational reformers had advanced the notion that universities should be established to meet the needs of the agricultural and industrial classes. Professor Jonathan Baldwin Turner of Illinois was the best-known advocate of a new type of agricultural and me-

School in Russia," *LP*, III (1889), 99; "The National Agricultural School at Douai, France," *LP*, XI (1893), 147–48.

33. U.S. Department of Agriculture, Division of Chemistry, *Record of Experiments . . . in the Manufacture of Sugar from Sorghum and Sugar Canes,* 37

34. "Schule fur Zuckerindustrie zu Braunschweig," *Deutche Zuckerindustrie,* XII (1887), 1567; *ibid.,* XIX (1894), 162; *ibid.,* XXI (1896), 184.

35. W. C. Stubbs, "Southern Agriculture," n.d. (Typescript in Stubbs Papers, Folder 14, SHC).

chanical education. In 1851 he asserted that "Work Alone Is Honorable," and therefore college students should be taught that manual labor has a dignity of its own.[36] Turner's views influenced the educational philosophy of agricultural and mechanical college administrators throughout the nineteenth century, and Stubbs's understanding of education was rooted in this tradition. Nevertheless, theory had a place in this practical plan of learning. For example, in the teaching of mechanics, reformers of the 1850s perceived that the discipline included "all applications of the principles of abstract mechanics to human art."[37] These educational ideas, which formed the basis of the 1862 Morrill Act, subsequently had a strong impact upon the nature of schools like Louisiana State University. An 1882 report by the university's board of supervisors stated:

> Our effort is to give the education contemplated in the Federal grant to the College; that is, to say, one both "liberal and practical." It is intended especially for those who expect to devote themselves to the industrial arts, or to agriculture, in Louisiana; . . . we assume as the special province of our college to prepare young men for the life of planter or plantation mechanic. No expensive laboratories, or costly farms are required to illustrate our theoretical training. The neighboring cotton fields, sugarhouses, oil mills, gins, etc., etc., conducted practically and for profit, are open to the study and inspection of the student, and, . . . will complete the labors of the lecture room.[38]

It is more than likely that Stubbs had been exposed to a similar ideology at the Alabama Agricultural and Mechanical College during his thirteen-year tenure there, and that he later attempted to apply these concepts at the sugar school.

The initial faculty consisted of five professors. Stubbs taught agricultural subjects, while H. E. L. Horton and J. T. Crawley instructed students in organic, agricultural, and analytical chemistry. In addition, station engineer Robert Burwell was responsible for teaching

36. Gould P. Colman, *Education and Agriculture: A History of the New York State College of Agriculture at Cornell University* (Ithaca, N.Y., 1963), 32.

37. Alfred Charles True, *A History of Agricultural Education in the United States, 1785–1925* (Washington, D.C., 1929).

38. Isaac Edwards Clarke, *Art and Industry: Education in the Industrial and Fine Arts in the United States: Industrial and Technical Training in Schools of Technology and in U.S. Land Grant Colleges* (Washington, D.C., 1898), 325.

courses in drawing and mechanical engineering, while a practical sugar maker directed students in various sugarhouse operations. Later, Beeson and Maxwell replaced Horton and Crawley on the staff. With Maxwell's departure in 1895, Stubbs hired Levi Wilkinson, a Heidelberg-educated chemist.

The student body was composed "for the most part, of men of maturity, from 25 to 40 years of age, who had previous sugar house work, and were anxious to supplement their experience with a certain amount of theory."[39] The two-year course consisted of lectures, recitations, and practical work, covering agriculture, qualitative and quantitative analysis, organic and sugar chemistry, physics, mechanics, applied mechanics, machines, boilers, mechanical drawing, the laws of crystallization, and the art of sugar making. The mornings consisted of work in the classroom, and in the afternoons students did practical work in the fields and sugarhouse. During the three-month harvest, students and professors ran the sugarhouse "in the experimental manufacture of sugar." Nevertheless, the conceptual foundations of the sugar trade were not neglected. The school bulletin stated that "while every manual act in the construction of machinery is part of their course, the scientific principles receive their mental attention."[40]

The basic structure and content of the sugar course changed little until after the turn of the century. Both chemistry and mechanical engineering were taught, but there is no evidence that their interrelationships were ever emphasized in the classroom, though they may well have been united in practice. For example, in the 1895 junior and senior examinations, rigid distinctions were made among questions in agriculture, mechanical engineering, and chemistry.[41] Only the sugar-making examinations required combined knowledge by asking the student to provide the necessary scientific principles underlying several discrete process operations—extraction, clarification, filtration, evaporation, and the use of the centrifuge. The student was required to explain the chemical and mechanical basis for the most efficient method of conducting each of these operations.

39. "The Audubon Park Sugar School," LP, X (1893), 385.
40. "Sugar School," LP, IX (1892), 98.
41. LP, XIV (1895), 403–405.

The American-born graduates of the Audubon Sugar School followed several distinct career paths. The institution was created to supply the Louisiana sugar industry with scientific experts, and a number of former students found employment at plantations owned by patrons of the experiment station.[42] Trosclair and Robichaux, Leonce Soniat, Andrew Gay, and Leon Godchaux all benefited from the services of Audubon graduates. A few of the students, like T. L. Schmidt, John Stone Ware, and E. C. Webre were sons of planters, and upon completion of their course, they assisted in the family business. In addition, many graduates were hired by the beet-sugar enterprises of Colorado and California, and by Hawaiian sugar manufacturers. Perhaps as a result of Maxwell's appointment as director of the Hawaiian Sugar Experiment Station, at least four former students were hired to perform various duties at the large sugar plantations located on the islands. However, not all of the graduates worked for others. Several were self-employed in Louisiana as consultants and businessmen, and a few eventually invested their savings from sugarhouse employment toward the purchase of their own plantations.

The 1896 depression in the Louisiana sugar industry, along with the outbreak of the Cuban civil war, effectively curtailed the Audubon Sugar School's educational activities. Cuban students had comprised a significant percentage of the student population; during 1894–1895, twelve of thirty-five students came from Cuba, but after the initiation of the rebellion against Spanish rule, only three of the twenty-three pupils enrolled were Cuban.[43] Thus, by the fall of 1896 the prospect for new students at the Audubon Sugar School appeared dim. The institution had never been a financial success. During 1895–1896 only $2,800 had been collected in enrollment fees, and Stubbs's 1896 recommendation to close the school came as no surprise to the LSPA.

42. For information on graduates, see "Audubon Sugar School," *LP,* XIII (1894), 10–15, XV (1895), 385–86, XVII (1896), 2, "Our Sugar School," *LP,* IX (1892), 95; "The Sugar School," *LP,* IX (1892), 313; "The Closing Exercises at the Sugar School," *LP,* XV (1895), 1–2.

43. William Carter Stubbs, "Audubon Sugar School," n.d. (Typescript in Stubbs Papers, Folder 10, SHC).

However, Stubbs's opinion was based on several considerations besides the tenuous financial condition of the school. Many of the students who had enrolled in the program at Audubon Park were poorly prepared to pursue the theoretical studies offered at the school. It was found that these individuals required an extra year of training. As a result, the faculty was burdened with additional teaching duties, which inevitably prevented them from conducting research. Stubbs had initially thought that students would assist in the research projects of the station, but in reality he found that "excessive teaching and superintendence of practical work has left but little time for experimental research."[44] The experience gained by the faculty at the school also made them attractive to other employers, and they were able to command higher salaries than had originally been anticipated. As illustrated by the cases of Horton, Crawley, and Maxwell, once a teacher gained knowledge through practice and experience at the station, he was offered positions with salaries that Stubbs could not match. In the early fall of 1896, Stubbs's recommendations to close the school were adopted by the planter executive committee responsible for station policy.

The Audubon Sugar School at Louisiana State University

Undaunted by the failure of the privately sponsored sugar school, Stubbs decided to campaign for the establishment of a second sugar course at Louisiana State University. In a November, 1896, letter to President T. D. Boyd, Stubbs expressed hope that the Audubon Sugar School curriculum would be incorporated into a four-year degree program at the university. The experiment station director felt that the primary purpose of such university training was to educate the sons of planters and to produce experts capable of assuming responsibility for the operations of large plantations or central factories. Stubbs proposed that the practical aspects of sugar manufacturing be taught during the junior and senior years. During these years the students would be in residence at the station during the fall harvest. There they would "be drilled in the planting and harvesting of cane,

44. Stubbs, undated address [1896?] (Typescript in Stubbs Papers, Folder 14, SHC).

effects of irrigation, effects of fertilizers, and be acquainted with the large number of varieties of cane . . . under cultivation."[45]

Stubbs was firm in stating to Boyd that he did not want the course to be reduced to what he considered a modern course. He remained tied to the educational philosophy responsible for the establishment of the nineteenth-century agricultural and mechanical colleges in America. In Stubbs's opinion, a successful educational program focused on practical experience gained through manual labor. But Stubbs acknowledged that because both theory and practice were crucial, instruction required a delicate balance of the two. His goal was "to make real experts, skilled in chemistry, mechanics, agriculture, sugar making and drawing."[46]

Boyd and the Louisiana State University Board of Supervisors felt that the Audubon Sugar School program would prove to be a valuable addition to the university's curricula, and the sugar school opened at Baton Rouge on September 15, 1897. The school's prospectus stated: "This course is especially rich in Chemistry and Mechanics, and is designed to train sugar experts—men who can perform scientifically and at the same time practically and economically all the operations incident to the cultivation and manufacture of sugar."[47] The sugar course became a five-year program at the university, enabling high-school graduates to have three years of preparation before entering the two-year professional course. During the fourth and fifth years, the students spent ten weeks each fall at the Audubon Park experiment station learning the practical operations of sugar processing, as well as participating in research projects conducted by the staff. In the engineering classroom they were trained in both theory and practice. The students graduated proficient in mechanical drawing, with a knowledge of machine shop and the ability to measure the mechanical parameters associated with pumps, wheels, gears, and

45. Stubbs to Thomas D. Boyd, November 11, 1896, in Stubbs Papers, LSU.

46. Stubbs to Boyd, November 11, 1896, in Stubbs Papers, LSU.

47. Louisiana State University Board of Supervisors, Report . . . to the General Assembly of the State of Louisiana for the Sessions 1896–97 and 1897–98 (Baton Rouge, 1898), 11. Histories of Louisiana State University and its presidents include: Marcus M. Wilkerson, Thomas Duckett Boyd (Baton Rouge, 1935); Germaine M. Reed, David French Boyd (Baton Rouge, 1977); Fleming, Louisiana State University, 1860–1896.

shafts. They were taught the fundamentals of thermodynamics, material science, and the kinematics of machinery. By 1900 engineering courses at Louisiana State University emphasized design based on the empirical determination of the capacity of an apparatus and its proper proportions.[48]

However, one student, Robert Glenk, felt that his training in theoretical mechanical engineering would prove to be of little benefit to him. In an 1898 letter to Stubbs, Glenk complained that his studies using George Holmes's *The Steam Engine* and his drawing of bolts, nuts, rivets, and engine cylinders in various positions was "apt to be quite awkward in the shop." Stubbs diplomatically replied to the disgruntled student that it was also necessary to acquire theoretical knowledge, and that theoretical mechanics was a glaring weakness of all of the previous sugar school graduates.[49]

In addition to the engineering courses taught to Audubon Sugar School students, the university's chemistry department offered instruction in a variety of technical and analytical chemistry subjects related to the large-scale manufacture of sugar. Between 1897 and 1905 Charles Edward Coates was professor of chemistry at Louisiana State University. Later he became dean of the Audubon Sugar School there. Coates was born in Baltimore in 1866 and graduated from Johns Hopkins University with a bachelor's degree in 1887.[50] After a year of graduate study at Johns Hopkins, Coates spent about sixteen months in Germany, first at Freiburg and then at Heidelberg. He then returned to Johns Hopkins, where he completed his doctoral dissertation in 1891.

Coates taught four advanced courses in chemistry to the students enrolled in the sugar program. A two-term course entitled "Technical Organic Chemistry" consisted of lectures in the principles of agricultural chemistry, methods of analyses approved by the Asso-

48. Louisiana State University, *Catalogue, 1899–1900,* 20, 41–42, 53–55.
49. Robert Glenk to Stubbs, June 13, 1898, in Glenk Papers, Manuscripts Department, Special Collections Division, Howard-Tilton Library, Tulane University. Stubbs to Glenk, June 14, 1898, in Glenk Papers, Tulane.
50. Charles Edward Coates, "The Action of Aniline and of Toluidine on Ortho-Sulphobenzoic Acid and its Chloride" (Ph.D. dissertation, Johns Hopkins University, 1891).

ciation of Official Agricultural Chemists, and a general discussion of technical organic chemistry. In the laboratory students analyzed sugars, fertilizers, and agricultural products, and they visited sugarhouses, gasworks, and cottonseed-oil works.[51] Seniors were required to enroll in an advanced organic chemistry course, where they studied the reactions of amides, albuminoids, and carbohydrates. In the laboratory students performed elemental organic analyses and examined various sugars by characterizing their optical and chemical properties. Coates's research efforts were primarily concerned with organic analysis.[52] Prior to his appointment as director of the Audubon Sugar School in 1905, Coates had pursued a variety of research topics, including the role of nitrogen in the growth of cotton and a characterization of the crude petroleum oils discovered in Louisiana. Perhaps because of increased administrative duties, Coates's publications after 1908 were not as technically oriented. Instead, they were usually general-interest, informative articles targeted for the large planter or central sugar factory owner.

Until Stubbs's retirement in 1905, agriculture remained a central component of the sugar course. Students were required to complete six courses in agriculture and two in botany during their five years of study. Freshmen studied general agriculture, a course based primarily on S. W. Johnson's *How Crops Grow* and *How Crops Feed*.[53] Lectures focused on the physical and chemical properties of soils, the process of nitrification, and proper drainage practices. In addition, students were taught cultivation and planting techniques, and

51. Louisiana State University, *Catalogue, 1900–1901*, 31.

52. Coates's publications include: Charles E. Coates and Alfred Best, "The Hydrocarbons in Louisiana Petroleum," *Journal of the American Chemical Society*, XXV (1903), 1153–58, XXVII (1905), 1317–21; Charles E. Coates, "The Sugar House Laboratory and Chemical Control," *LP*, XLIX (1912), 92; "The Equipment of a Sugar House Laboratory," *LP*, XLIX (1912), 160–61, L (1913), 174–76; "The Adeline Sugar Company," *LP*, L (1913), 28–31; "White Sugar in Louisiana Gramercy Central Factory and Sugar Refinery," *LP*, L (1913), 139–41; "White Sugar Making in Louisiana," *LP*, LIV (1915), 12; "The Making of Sugar in Louisiana with Special Reference to White Sugar," *LP*, LV (1915), 157–59, LVI (1916), 60–62, LVII (1916), 172–73; Charles E. Coates and L. C. Slater, "A Study of the Sirup Precipitate in White Sugar Manufacture," *Journal of Industrial and Engineering Chemistry*, VIII (1916), 789–92.

53. Louisiana State University, *Catalogue, 1902–1903*, 26–27.

thus learned about tillage, crop rotation, the preparation of organic fertilizers, and the application of various types of manures and commercial fertilizers to the soil.

However, the Audubon Sugar School's program of agricultural instruction was much broader than necessary for the effective instruction of sugar plantation managers. For example, students were required to complete courses in stock breeding and stock feeding, receiving instruction in the raising of dairy and beef cattle, hogs, horses, and mules. They also heard lectures on the operation of creameries and cheese factories. During the fifth year, students took two final courses in the agriculture of sugar in which they studied farming methods for producing maximum sugar tonnage as well as maximum sugar content from cane, beets, and sorghum.

Enrollment at the Audubon Sugar School fluctuated between ten and twenty-four students annually between 1900 and 1905. Although printed records concerning the career paths of the school's graduates are incomplete, it appears that many of the former students found employment outside Louisiana. With the rapid decline of the Louisiana sugar industry after 1903, and the fragmentation of planter interests in Louisiana, the primary objective of the Louisiana State University Audubon Sugar School was no longer to train planters' sons for future employment in that state. Rather, it now trained experts for the international sugar industry, particularly Cuba's large central sugar factories.

Sugar Engineering at Tulane University

Louisiana State University was not the only Louisiana university to offer a program in sugar chemistry and sugar engineering at the turn of the century. Through the efforts of Levi Wilkinson, elements of the sugar course first taught at Audubon Park were transferred to Tulane University in 1896. This private university was legally chartered in 1884, the result of a generous donation from New Jersey merchant Paul Tulane, a former New Orleans resident.[54] Tulane entrusted the details of founding this educational institution to Randall Lee Gibson, LSPA member and United States senator, who as-

54. For a good history of Tulane University, see Dyer, *Tulane: The Biography of a University, 1834–1965.*

sembled a group of prominent Orleanians to act as trustees for the new university. In his letter to Gibson and the other administrators, Tulane stated: "By the term education, I mean to foster such a course of intellectual development as to be useful and of solid worth, and not merely ornamental or superficial. I mean you should adopt the course . . . conducive to immediate practical benefit, rather than theoretical possible advantage."[55] Reflecting the basic concerns of its benefactor, the school's early curriculum concentrated on practical technical training. John Morse Ordway, formerly of the Massachusetts Institute of Technology, was hired in 1884 to organize its program in manual training and was concurrently appointed professor of applied chemistry and biology. During the 1880s the majority of the Tulane students were enrolled in courses such as wood shop, metal working, and practical machine building.

Born in 1823 in Amesbury, Massachusetts, Ordway was apprenticed to a chemist and apothecary in the nearby town of Lowell at age thirteen. After graduating from Dartmouth College in 1844, he entered the chemical manufacturing business. Later he became superintendent of the Roxbury Color and Chemical Works, and in 1850 he moved west, where he was appointed professor of chemistry at Grand River College in Trenton, Missouri. Ordway returned to Roxbury in 1854, and four years later he accepted a position with the Hughesdale Chemical Works of Johnston, Rhode Island. During the 1860s he was employed as a chemist and superintendent of a Manchester, New Hampshire, print works and later as a chemist for the Bayside Alkali Works in south Boston. In 1869 he became professor of industrial chemistry and metallurgy at Massachusetts Institute of Technology, where in 1878 he was elected chairman of a faculty committee responsible for the direction of the School of Mechanic Arts.[56]

55. Paul Tulane to Randall Lee Gibson and others, May 2, 1882, quoted in Clarke, *Art and Industry*, 80.

56. "John Morse Ordway," in John Morse Ordway Papers, Folder 1, Manuscripts Department, Special Collections Division, Howard-Tilton Library, Tulane University; Anna Ashbey, "Curriculum Vitae for John Morse Ordway, 1823–1909," in Ordway Folder, University Archives, Special Collections Division, Howard-Tilton Library, Tulane University. Ordway's publications include: "Waterglass," *American Journal of Science*, 2nd ser., XXXII (1861), 153–65, XXXIII (1862), 27–36, XXXV (1863), 185–96, XL (1865), 173–90; "Nitrates of Iron," *ibid.*, 2nd ser., XL (1865), 316–31. Ordway's

Ordway's involvement in manual training at Massachusetts Institute of Technology was probably the result of the efforts of John D. Runkle, the school's president. Runkle, who had viewed a Russian exhibit at the 1876 Philadelphia Centennial Exposition on the theory and practice of tool instruction, became convinced that education in woodworking, forging, drawing, pattern making, and machine shop would be of value in mechanical engineering studies as well as in general education.[57] Thus, during the late 1870s and early 1880s, other universities followed the institute's example, and manual training became widely adopted in the curricula of technical schools, private universities, and land-grant colleges.[58]

Ordway's broad industrial experience in part shaped the nature of chemical instruction at Tulane during the 1880s. Students in the natural science and mathematics programs at the school normally enrolled in courses in organic and inorganic chemistry, qualitative and quantitative analysis, and industrial chemistry. Ordway's lectures were not only illustrated by experiments, but also characterized by "problems on the cost of manufacture and the qualitative relations of products to the materials from which they are made." Classroom topics included the reduction of ores, the chemistry of photography, the processes of dyeing and bleaching, and the manufacture of iron, steel, explosives, and soap. Apparently, Ordway paid particular attention to problems in sugar manufacturing in both his lectures and the required laboratory. Furthermore, the 1885 Tulane catalog mentioned that "those who desire it may give some attention to sugar testing and analysis, both by polariscopic and chemical methods."[59]

William Preston Johnston, the university's first president, recog-

---

activities at MIT are mentioned in Samuel C. Prescott, *When M.I.T. Was "Boston Tech"* (Cambridge, Mass., 1954, 114, 117–18.

57. For further information on Runkle, Ordway, and manual training, see C. M. Woodward, *The Manual Training School, Comprising a Full Statement of Its Aims, Methods and Results* (Boston, 1887); Charles H. Ham, *Manual Training: The Solution of Social and Industrial Problems* (New York, 1886).

58. Clarke, *Art and Industry*, 87.

59. Tulane University, *Catalogue of the Academical Department, 1884–85* (New Orleans, 1885), 57–58.

nized that if the institution was to succeed, it would have to respond to local economic conditions. Seeing the opportunity to garner the support of private interests, Johnston wrote LSPA secretary J. Y. Gilmore in 1887 detailing Tulane's programs for training students in the practical arts, chemistry, and mechanical engineering. He stated: "We are quite prepared to carry this broad scientific development of our students into any special lines of training for particular industries, whenever any public spirited individual or association shall furnish the means for such a purpose."[60]

However, since the LSPA ignored Johnston's implicit suggestion that sugar engineers could be educated at Tulane if given industry support, Ordway continued to develop a broad program in applied chemistry. The 1890–1891 Tulane *Register* mentioned Ordway's willingness to train engineering and graduate students in aspects of the chemical arts, particularly in "investigations . . . having reference to economy and excellence of production." In the 1894–1895 school year Tulane offered a course in chemical engineering for the first time. It was claimed that the program was initiated in response to the changing industrial environment of south Louisiana, and its purpose was "to train young men who should be especially well qualified to undertake the direction of manufacture of sugar, of cotton, and other textile fabrics, of oils, glycerine, soaps, paper, alcohol, fertilizers."[61]

The chemical engineering curriculum at Tulane included instruction in chemistry, mathematics, physics, and engineering. During the freshman and sophomore years students were taught various shop skills. Juniors were exposed to both physics and industrial chemistry, while in the senior year students concentrated on mechanical engineering and received additional course work in industrial chemistry. In addition to undergraduate instruction, Ordway developed an intensive graduate course in chemical engineering. The advanced student was given the choice of concentrating on several narrowly defined subjects in chemical technology: sugar, starch, and

60. Louisiana Sugar Planters' Association, *Regular Monthly Meeting, December, 1887,* 4.

61. Tulane University, *Register, 1890–91,* 48; Tulane University, *Register, 1894–95,* 66.

glucose; textile manufactures; metallurgy; oils and soaps; manufacture of alkalis and salts; pottery, glass, cements; electrochemistry; sanitary chemistry; tanning; India rubber; and paints.

Ordway felt that chemical engineering was a completely different discipline from both chemistry and mechanical engineering. While the chemist studied chemical reactions in the laboratory, the chemical engineer improved the efficiency of large-scale processes in the plant by introducing new designs of complex mechanical apparatuses. Thus the chemist and chemical engineer had different functions and therefore should be assigned different responsibilities. The practical objectives of this new professional demanded the extensive use of mathematics as well as the general laws of physics. Theoretical knowledge, coupled with careful planning and management, enabled the chemical engineer to devise new apparatuses and optimize their performance. However, the chemical engineer did not simply apply the principles of mechanical engineering to problems in chemical technology. The design of chemical apparatuses was not merely based on principles in machine design.[62] Ordway clearly perceived that the highly corrosive environment associated with reaction vessels and the complex nature of chemical reactions posed a set of unique problems for the chemical engineer. Thus, one could not compare the scientific and technical requirements necessary for the design of a machine to stamp metal parts with those necessary to construct a high-pressure, high-temperature reaction autoclave.

Ordway viewed chemical engineering education as a balance between courses based on chemical knowledge (such as general chemistry, analytical chemistry, industrial chemistry, electrochemistry, metallurgy, and mineralogy) and instruction in pure mathematics, analytical mechanics, hydromechanics, thermodynamics, physics, drawing, and apparatus design. Both the design of process equipment and the direction of its efficient operation required the union of these disparate bodies of knowledge.

Led by Dean Brown Ayres of the College of Technology, Tulane administrators tried throughout the 1890s to establish a course in sugar engineering that would complement Ordway's chemical engi-

62. John M. Ordway, "Chemical Engineering," *Proceedings of the Society for the Promotion of Engineering Education*, V (1898), 187–99.

neering program. Ayres, with a doctorate in physics from the Stevens Institute, was professor of physics and astronomy at Tulane between 1880 and 1904 and dean of the College of Technology from 1894 to 1900.[63] He recognized that the South was experiencing an industrial revolution, and that manufacturing would soon become as important as agriculture in the region. Ayres anticipated that industrial growth and the modernization of existing organic-extractive trades would result in an increased demand for engineers possessing both theoretical and practical knowledge. He also realized that if Tulane responded to these structural changes by modifying its course offerings in engineering, it would most certainly benefit by increased enrollment and by the receipt of more operating funds. In 1892 Ayres outlined his views to President Johnston.

> Louisiana is really a manufacturing state. . . . We should supply—in fact, anticipate—this demand, and should be in a position to turn out trained *sugar engineers* and cotton and oil engineers, that would be competent to so manage and improve processes that the commercial results as well as quality of product would be improved, I believe that all this is within our reach, and when we are in a position to offer to the people of this state thoroughly well worked out courses of this sort for their sons, . . . [these] will command their support and approval.[64]

Ayres hoped that an emphasis on sugar engineering within the newly established chemical engineering program would increase the popularity of the chemical technology curriculum at Tulane. Perhaps Ayres was searching for a way to duplicate the successful electrical engineering department at Tulane in other areas of the school's College of Technology. With the temporary closing of the Audubon Sugar School, Ayres felt that a void in technical education existed in Louisiana and particularly in New Orleans. He felt that Tulane would profit from the establishment of a course in technology in a state economically dominated by sugar interests.[65]

63. "Professor Brown Ayres, Ph.D." (Typescript in Ayres Folder, University Archives, Howard-Tilton Library, Tulane University).

64. Brown Ayres to William Preston Johnston, June 7, 1892, in Ayres Folder, Tulane.

65. New Orleans *Daily Picayune*, September 29, 1896 (Typescript in Audubon Sugar School Folder, University Archives, Howard-Tilton Library, Tulane University.

Levi Wilkinson, a professor of organic and sugar chemistry at the Audubon Sugar School between 1895 and 1896, directed studies in sugar technology at Tulane. Born in Alabama in 1860, Wilkinson had received his bachelor and master of science degrees from the Alabama Agricultural and Mechanical College at Auburn.[66] Between 1893 and 1895 he pursued an advanced degree at Heidelberg. After receiving his doctorate, he accepted a position at the Audubon Sugar School and concurrently cultivated relationships with Tulane students and faculty members. When Stubbs decided to close the school at Audubon Park, Wilkinson received permission to use Tulane's classroom and laboratory facilities to teach sugar chemistry fundamentals to a group of special students. The special students averaged twenty-six years of age and had diverse backgrounds: two had already earned four-year degrees, and previous occupations of others included chemical manufacturer, teacher, bookkeeper, business school graduate, and traveling salesman.[67] Even though the university received valuable publicity, the administration made no commitment to sponsor the program for more than one year.

Dean Ayres strongly supported Wilkinson's work at Tulane, and in 1897 they both actively campaigned for the establishment of a permanent sugar engineering course. While Wilkinson secured necessary equipment, Ayres attempted to convince President Johnston that sugar engineering would complement the existing programs in the physics and mechanical engineering departments and more fully utilize the university's physical plant. In formulating course requirements, Ayres felt that Tulane would develop a sugar engineering course that concentrated on the chemical and mechanical principles

66. "L. W. Wilkinson," February 11, 1901 (Typescript in Wilkinson Folder, Tulane). Wilkinson's publications include: "Proposed Method for the Analyses of Native Phosphates Containing Iron from Aluminum," *Proceedings of the Association of Official Agricultural Chemists*, VIII (1891), 107; "Proposed Change in the Methods for Estimating Water Soluble and Acid Soluble Phosphoric Acid," *ibid.*, VIII (1891), 108; "Uber die Jodoniumbasen aus p-Chlorjodbenzal," *Berichte der Deutschen Chemischen Gesellschaft*, XXVIII (1895), 99–101; "The Absorption of Sucrose by Bone Black," *LP*, XVI (1896), 390–91. Victor L. Roy Diary, April 7, 1895 (MS in Victor L. Roy Papers, Special Collections, Hill Memorial Library, LSU).

67. Brown Ayres to William Preston Johnston, January 18, 1897, in Sugar Engineering Folder, University Archives, Howard-Tilton Library, Tulane University.

of the discipline, leaving the agricultural aspects of sugar manufacturing for Louisiana State University.[68]

Ayres's ideas concerning the new sugar engineering program at Tulane reflected changes that were taking place in the Louisiana sugar industry during the 1890s. The traditional plantation culture, in which the planter was responsible for both growing the sugarcane and manufacturing raw sugar, was being supplanted by a new order. Central factories, located primarily on the largest plantations in Louisiana, processed not only the plantation owner's crop, but also sugarcane purchased from small farmers nearby. This increase in process capacity was necessary to justify the capital expenditures for advanced technology and machinery. Ayres saw the need for a chemical and engineering expert to control and supervise these plants. However, these experts would not be required to possess a knowledge of agricultural chemistry, since problems dealing with the growth of the cane were divorced from those of manufacturing.

Ayres's arguments convinced the Tulane University Board of Administrators, and in June, 1897, they authorized the establishment of a course in sugar engineering for one year, with Wilkinson receiving a salary of $1,000 in addition to the fees from special students. Also, $250 was appropriated for the purchase of apparatuses that would aid instruction in practical plant operation. The purpose of the new course was clearly stated in a newspaper article.

> The plan of the course which we have drawn up includes only so much of the agricultural side of the question as is necessary to give the student a fair idea of the raw material with which he is to deal. Our object will be to educate the sugar engineer, rather than the sugar planter. By the term "sugar engineer" we understand one who is competent to direct every part of the manufacture of sugar from cane or beets, and to do this in the most accurate and economical manner. Such a man must be a combination of a mechanical engineer, electrical engineer and an industrial chemist.[69]

The sugar engineering program united the aspects of sugar chemistry with principles governing the design, proper erection, and eco-

---

68. Ayres to Johnston, April 7, 1897, in Sugar Engineering Folder, Tulane.
69. New Orleans *Times-Democrat*, July 17, 1897 (Typescript copy in Sugar Engineering Folder, Tulane).

nomical management of the sugar factory and its machinery. Thus, the student would be familiar not only with the processes of clarification and procedures of chemical control, but also with the installation, repair, and proper operation of boilers, pumps, engines, dynamos, and centrifugals.

Sugar engineering at Tulane was quite similar to the course in chemical engineering, as both programs focused on mechanical engineering.[70] In fact, the first two years of the curriculum were the same as any other engineering course at Tulane. Instruction was given in general shop techniques, as well as drawing, basic mathematics, and physics. Practice work in the various shops served as an integral component of the student's training during the freshman, sophomore, and first part of the junior year. The students learned carpentry through a series of exercises in planing, sawing, rabbeting, plowing, paneling, splicing, mortising, tenoning, and dovetailing. Wood turning and patternmaking involved practice in turning cylindrical, spherical, and other curved surfaces. Students made patterns and cores for columns, pipes, pipe joints, pulleys, and gears. Forging instruction included the tasks of heating, bending, drawing, welding, annealing, and case hardening of iron, as well as the making and tempering of steel punches, chisels, machine cutting tools, and springs. Machine work in metals included exercises in turning, planing, slotting, drilling, thread cutting, and pipe cutting. Students gained foundry experience in greensand molding, core making, and the pouring of castings. This practical experience enabled students to make various equipment in the university shop, including steam engines, oil-testing machines, balance beams, crane brackets, grate bars, floor plates, and engine brakes. Further, they learned to repair rack teeth and dynamo castings. In a very real sense, the sugar and chemical engineering programs at Tulane fused the nineteenth-century foundry tradition with scientific knowledge in the form of instruction in chemistry, physics, and theoretical mechanics.

This emphasis on practice was continued in the mechanical en-

70. Tulane University, College of Technology, *Course in Sugar Engineering and Sugar Chemistry, Circular of Information* (New Orleans, 1899); "Course in Sugar Engineering at Tulane," n.d. (MS in Audubon Sugar School Folder, Tulane).

gineering and sugar chemistry laboratories. In the course on the strength of materials, normally taught during the junior year, students tested boiler steel, the deflection of beams, and the strength of columns. They also calibrated heavy springs and determined the compressibility of cast iron, brass, bricks, marble, and cements. Sugar and chemical engineering students determined the densities, viscosities, and flash points of oils, and examined their lubricating qualities utilizing an oil-testing machine. In steam engine tests, the students calibrated indicators, steam and vacuum gauges, thermometers, and planimeters. Using these instruments, they empirically obtained the correct valve settings and determined water consumption, effective horsepower, and efficiency. Steam boiler testing included the establishment of the capacity and efficiency of boilers, analysis of flue gases, and practical experience in their operation and repair. Machine design was also an important component of the curricula. Single, compound, and triple-expansion boilers were drawn, as well as Corliss and automatic engines, sugar mills, centrifugals, vacuum pans, and multiple-effect evaporators. Basic knowledge of machine elements—namely spur and bevel gearing, cams to convey motion at various speeds and in various directions, pulleys, epicyclic trains, and clock mechanisms—were the building blocks for students designing complex mechanical and chemical apparatuses.

Practical work in the sugar chemistry laboratory included the study of the characteristic reactions of alcohols, aldehydes, acids, ketones, ethers, and amines. Also, students performed analyses of fertilizers, soils, feed stuffs, fuels, and residual sugarcane ash. In addition to polariscopic analyses of carbohydrates, the reaction of these compounds with acids, bases, and oxidizing agents was observed. Wilkinson directed a special study of alcoholic, acetic, lactic, and butyric fermentations of sugars and other natural products. Processes of clarification—such as carbonatation, sulfuring, liming, the lead sucrate process, the strontium process, and electrolytic processes—were duplicated on a small scale in the laboratory. Students isolated sucrose, lactose, maltose, raffinose, dextrose, dextrin, levulose, and galactose for polariscopic evaluation. They also practiced analyzing unknown synthetic mixtures of the above carbohydrates and common sugarhouse raw materials and products—cane juices, syrups, massecuite, sugars, molasses, glucose syrup, sugarcane, bagasse, and beets.

Throughout the 1890s, practice was the most important component of the curriculum. This emphasis upon experiential knowledge may well have been a reflection of the demands of potential employers. In·1897 a course prospectus stated: "Thorough recognition is given to the fact that graduates must be practical men. By this is meant capability in handling personally large machinery and such operations. Such experience is first given in the well equipped shops of the University, then in handling the engines, pumps and boilers of the University and the various power plants of the city."[71] In addition to the education received on campus, students conducted engineering investigations at the Jordan Avenue Pumping Station in New Orleans, the Ernst Rice Mill, and the Hennes Steam and Power Plant at Tulane. To expand the students' specific knowledge of sugar making, the university sponsored several visits to the plantation of Frank Ames, a graduate of the Massachusetts Institute of Technology. The students resided at the plantation for several days, observing the processes and receiving instruction from practical sugar boilers in the operation of the pan, the multiple-effect evaporator, and clarifiers.

It is clear that the aim of the Tulane sugar course was to produce a graduate possessing "one-man power." Thus, on the surface, the ideas of Tulane educators seemed to resemble those of planter John Dymond. However, while Dymond was interested in promoting a program that educated the sons of planters for survival in a new international sugar market, Tulane's course was more closely aligned with the shop culture of the foundry and reflected northern industrial interests. An 1897 description of the sugar engineering course stated:

> These sugar houses must be under the control of one man so that there may be undivided authority and the consequent responsibility. That this may be so, this man must be both an engineer and chemist. The aim then of the Sugar Engineering course will be to produce men capable of taking sole control of a sugar house in the fields of electrical, mechanical and chemical engineering. Above all, however, it is realized that the men must be practical as well as theoretical in each of these departments. . . .

71. Tulane University, *Course in Sugar Engineering*, 8.

The very great value of the systematic courses and training that are already given in the university will be better appreciated, perhaps, if attention is called to the fact that the designing and construction of sugar cane mills is simply a special application of the principles of machine design and taught in the senior year. Similarly, the design and construction of multiple effects, vacuum pans, centrifugals and pumps are special applications of the principles of thermodynamics and steam engine design. In other words, as far as the mechanical and electrical sides are concerned, the sugar engineering student will find he has only to apply old principles in a new field.[72]

Sugar engineering at Tulane was similar to the program in chemical engineering in that chemistry and mechanics were unified by practice. The design and construction of chemical apparatuses, activities closely associated with the foundry tradition, were the key elements of the sugar engineering program. In these courses, theory could be meaningfully applied and directed toward the solution of problems associated with large-scale sugar manufacturing.

By early 1899 Wilkinson began expressing his displeasure with his duties at Tulane. At that time, the sugar engineering program had expanded to eleven students, including three graduate students. Wilkinson's time was almost entirely devoted to directing the day-to-day affairs of the laboratory. He bitterly complained that though he had the professorial title, his salary was not on the level of other professors.[73] In addition, his teaching was severely hindered by opposition from Tulane's professor of general chemistry, Dr. C. P. Caldwell. Wilkinson detailed his problems in a 1901 letter to Tulane's newly installed president, E. A. Alderman.

> Another difficulty in the department of chemistry is *the fact that the professor of General Chemistry fails to see or appreciate the importance of Industrial or Sugar Chemistry.* In the early part of my connection with the University I was told by him in unmistakable terms that the University had no place for my work. As recent as last session he asked me what need Tulane had for a chair of Industrial Chemistry. . . . These remarks show how unacquainted he is with the real conditions of our immediate surroundings and that he utterly fails to appreciate the great importance

72. "Sugar Engineering in Tulane University" [1897?] (Typescript in Sugar Engineering Folder, Tulane).
73. Wilkinson to Johnston, March 10, 1899, in Wilkinson Folder, Tulane.

of the application of chemistry to the development of our agricultural and natural resources. . . .

Furthermore, a very serious difficulty now in chemical department is a lack of co-operation in and co-ordination of the work. . . . Now I do not say that the professor should dictate to or personally direct the work of Industrial and Sugar Chemistry, but he should manifest some interest in all departments of Chemistry and seek to develop them according to their respective importance.[74]

Wilkinson found an ally in President Alderman, who was firmly convinced that technical rather than classical studies were necessary to satisfy the economic and educational demands of the region. In Alderman's opinion, the South was shifting from an agricultural to an industrial economy, and "a few thousand dollars spent on the department of sugar engineering would enable it to multiply vastly ·the power of the sugar planters in Louisiana, and to affect the sugar industry of Latin America and Cuba. Is not Tulane University the logical place to disseminate the best knowledge about sugar manufacture? Indeed, the world has a right to demand this of us."[75]

However, Wilkinson proved incapable of formulating a true course in sugar engineering. His understanding of chemical technology was largely based on common technical textbooks of the period. His course in industrial chemistry, as described in the university's *Register*, was derived entirely from Frank Hall Thorp's *Outlines of Industrial Chemistry*.[76] Wilkinson began with a brief discussion of basic principles like evaporation, distillation, sublimation, filtration, crystallization, calcination, and refrigeration. He then reviewed the chemical reactions and machinery associated with various industries, including the manufacture of acids, alkalis, chlorine, bleaching powders, fertilizers, cements, glass, ceramics, pigments, sulfates, cyanides, alcohols, illuminating gas, coal-tar derivatives, soap, glycerin, sugars, starch, explosives, paper, dyes, fermentation, leather tanning, and the refining of numerous types of oils.

74. Wilkinson to Edwin Anderson Alderman, December 3, 1901, in Wilkinson Folder, Tulane.

75. Edwin Anderson Alderman, *Inaugural Address*, March 12, 1901 (Boston, 1904), 17.

76. Frank Hall Thorp, *Outlines of Industrial Chemistry* (New York, 1898).

Wilkinson's research never examined the interrelationship between the design of chemical apparatuses and the nature of large-scale chemical reactions. His primary investigative accomplishment was the examination of the formation of invert sugars, particularly dextrose and levulose, in open kettles. In fact, he remained tied to a traditional view of chemical education. The unification of chemistry and mechanical engineering ultimately became the responsibility of Tulane's engineering courses in the practical construction of sugarhouse apparatuses.

Prior to 1905 both Louisiana State University and Tulane University sponsored programs in sugar chemistry and sugar technology. Even though these programs were rooted in the theory of practical education emphasizing the dignity of manual labor, Louisiana State University's sugar curriculum concentrated on agricultural studies, while Tulane's focused on shopwork. The Audubon Sugar School at Louisiana State University had the ostensible purpose of educating the sons of planters; Tulane's course was shaped by the administrator's and the faculty's perceptions of Louisiana as an industrial state. The Audubon Sugar School program was formulated during a period when both cane growing and sugar manufacturing were closely associated with plantation operations. After 1890 these activities bifurcated, and while the Audubon school's curriculum reflected the old order, Tulane University structured a sugar engineering course that was directed at training technical experts for the industrial rather than the agricultural side of the rapidly modernizing sugar industry in Louisiana. Both of these schools' programs were formulated on the assumption that the Louisiana sugar industry would maintain its competitive position within the international sugar trade.

However, Louisiana sugar interests declined dramatically after 1903 as a result of falling prices, unfavorable tariff treaties, floods, and crop freezes. In addition, the once powerful LSPA fragmented and was unable to effectively lobby for planter interests in Washington. After 1900 the scientific institutions originally established by the LSPA fell under the control of the state government, which often had completely different objectives.

11

The Period of Decline

The Louisiana Sugar Planters' Association was one generation's response to increased foreign competition. And the LSPA proved to be the central organization in the scientific and technical infrastructure that emerged during the late nineteenth century in Louisiana. From this planters' group came first a sugar experiment station and then a unique school uniting theory and practice. This organizational network and its ability to serve the local sugar industry was only as strong as the parent organization. With the decline of the LSPA after 1900, however, these satellite institutions were drawn toward other institutional patrons with different interests and objectives.

Between 1885 and 1900 the research program at the experiment station, though formulated by station director W. C. Stubbs, was a direct response to the perceived needs of the Louisiana planters. Many of the station's projects began as discussion topics at LSPA meetings. For example, there were numerous debates involving both planters and foundrymen over the relative merits of diffusion and milling processes. The controversy was finally decided after a seven-year study at the Audubon Park sugar experiment station. Once Stubbs was elected to the LSPA executive committee in 1893, he was in a most favorable position to understand the demands of the association's leadership and acted accordingly. Station staff pursued scientific investigations on irrigation and drainage, soil mechanics and levee construction, the use of bagasse in papermaking, and the nature of the cane-borer insects only after the LSPA had defined these economically relevant problems. Furthermore, the Audubon Sugar School was established in response to local demands, and its stated purpose was to educate the sons of Louisiana planters. The peculiar combination of practical and theoretical training in agriculture, mechanics, and chemistry was an attempt by Stubbs and his station staff to develop a particular kind of sugar expert. Not only would this individual be able to assume a position of authority over the skilled sugarhouse artisans, but he would also possess the agricultural knowledge necessary for the cultivation of high yields of sugar-rich cane. Thus, the Audubon Sugar School's early curriculum was a reflection of the plantation culture of Louisiana.

During the 1890s the LSPA found its position within the planter community threatened by political turmoil and economic changes. With the subsequent fragmentation of political consensus within the LSPA, not only the quality and relevance of original research at the experiment station declined, but also administrators at Louisiana State University modified the content of the sugar course originally taught at Audubon Park.

The Fragmentation of the LSPA

By 1894 Democrats in the Cleveland administration had decided that the sugar bounty should be repealed and that sugar should be placed on the free-duty list. Between 1884 and 1892 the Democratic party had gradually shifted its position on sugar duties from a policy favoring tariff protection as a means to produce revenue to one advocating a "free breakfast table." Louisiana Democrats were in a tenuous position, since they could no longer count on partisan political support. Because of their previous unpopular stance on the silver question (the Louisiana congressional delegation had voted to stop silver coinage in favor of converting to a gold standard), their subsequent appeal for protection was ignored by fellow southern Democrats.[1]

_____

1. G. L. Pugh to William P. Miles, February 2, 1894, in Miles Papers, Box 5, Folder 70, SHC.

During the spring of 1894, an LSPA committee consisting of John Dymond, Henry McCall, James A. Ware, and Dudley Avery worked in Washington in an attempt to influence congressional members to vote against the proposed Wilson bill, which would drastically reduce sugar import duties. Expressing their opinion in a public letter, the committee asserted "that any attempt on our part to cast our lot with the Republicans for the purpose of defeating the Wilson bill, with the hope of a retention of the bounty, is impracticable and fraught with the greatest danger."[2] While the committee worked on tariff legislation in Washington, a movement outside the LSPA, led by Republican planters, called for a convention of sugar planters to meet at the New Orleans Academy of Music on May 11, 1894. At that gathering, former governor Warmoth and Republican sugar planter Marcus Parkinson attempted to persuade the large group of predominantly Democratic planters to reconsider their party affiliation.[3] However, the anticipated bolt of planters from the Democratic party did not occur.

Planter discontent with Democratic party policies erupted on September 4, 1894, when Ascension Parish planters met to discuss joining the Republican party *en masse*. Henry McCall, a prominent leader among planters, stated: "[We] must nominate a candidate thoroughly pledged to be an independent or a National Republican! Let no fears of the past deter us from action. The old war and reconstruction issues are dead; the force bill and election laws are buried never to be resurrected; the issues of the future are to be industrial, economical and social."[4]

On September 6 planters met at the Hotel Royal to plan an effective political strategy, for the movement to carry planters into the Republican party had gained considerable momentum. However, John Dymond was one of the few planters who vowed to maintain ties with the Democrats, and his remark that "the party is greater than the platform" undoubtedly cost him many friends. With that

2. William Porcher Miles Diary, a letter to the public from John Dymond, Henry McCall, James A. Ware, and Dudley Avery, May 4, 1894 (MS in Miles Papers, SHC), XXVIII, 103.

3. [William Porcher Miles], *Some Views on Sugar: By an Old-Fashioned Democrat* (N.p. [1894]).

4. "The Donaldson Meeting of Sugar Planters," *LP*, XIII (1894), 159.

statement Dymond was "literally snowed under," as indicated by the comments of LSPA leader H. P. Kernochan: "It is now clear that the time has arrived when we have to take our place either in the ranks of protection or remain in the party which is irrevocably committed to free trade. Considering our interests, the interests of the people of our State and of the whole country, our duty is only too plain. We have not abandoned the Democratic party—it has abandoned us."[5] A set of resolutions condemning the action of the Democratic Congress was adopted by the assembly, and another mass meeting was scheduled for September 17 in New Orleans to consider the entire problem.[6] At this meeting, the breach between the planters and the Democrats was completed. Kernochan, the temporary chairman of the convention, stated: "If we send without a struggle six Democrats to Congress from Louisiana this fall our doom is sealed, and we deserve it. It will not do us any good to send protection Democrats. They will only be curiosities. They will amount to nothing. We must come out straight and nominate our candidates on the National Republican ticket and do our level best to elect them, and if we succeed in electing them and carry our districts for protection our future is safe."[7]

Dymond considered these current political events to be the result of the emergence of a new order in the South. Southern politicians who failed to support the Louisiana sugar industry had not fought in the Civil War, but were "new men who have come to the front in the South during recent years and who have no feeling in common with their brethren of the other Southern states."[8] Dymond advocated that planters remain patient with the Democratic party and, in turn, for state Democratic party leaders to be sympathetic with the plight of the planter. For some reason, Dymond failed to realize that throughout the nineteenth century the sugar bowl had been socially, economically, and politically a distinct region. Many of Louisiana's elite sugar planters had been Whigs before the Civil War and later had Republican party affiliations, largely because they felt that protective tariffs were necessary for their economic well-being. Perhaps

---

5. "Planters Meeting at the Hotel Royal," *LP*, XIII (1894), 153–56.
6. Sitterson, *Sugar Country*, 337.
7. "Address of Hon. H. B. Kernochan, Temporary Chairman," *LP*, XIII (1894), 205.
8. "The Disintegration of the Solid South," *LP*, XIII (1894), 177.

Dymond was inflexible in his support of the Democratic party because he was a carpetbagger striving for social acceptability. But for many other planters, particularly those of southern backgrounds like Henry McCall, the Democratic party had nothing concrete to offer them.

The crisis over the Wilson bill effectively fragmented the LSPA. Thereafter, Dymond, as LSPA president, refused to involve the association in partisan politics, and as a result, a second planters' organization, the American Cane Growers' Association, was established in 1896. This group, claiming to represent all American sugar planters—cane and beet—had as its purpose political lobbying for securing favorable tariff legislation.[9] Although several members continued to participate in LSPA activities, the leadership of the newly formed association was composed of individuals who were concurrently attempting to build a powerful Republican organization in Louisiana—Henry McCall, B. A. Oxnard, William J. Behan, Jules M. Burguieres, D. D. Colcock, H. C. Warmoth, J. N. Pharr, and H. C. Minor.[10]

By early 1897 the LSPA was confronted with several crucial issues. To begin with, Dymond resigned as the organization's president, since he felt that his unpopular political position and his policy of nonpolitical involvement had hurt the organization.[11] However, a successor was not immediately found. Henry McCall refused a nomination to the presidency, and between March and May of 1897 a handful of faithful members attempted to select a new group of leaders. Finally, Emile Rost, an old friend of Duncan Kenner, agreed to become the association's president.

However, the LSPA never regained its former status among the Louisiana sugar planters. Unlike the LSPA, the American Cane Growers' Association was not reluctant to firmly ally itself with

9. "The American Cane Growers' Association of the United States," *LP*, XVII (1896), 33–34; "The American Sugar Growers' Society," *LP*, XVIII (1897), 55.

10. D. D. Colcock to W. P. Miles, December 18, 1896, in Miles Papers, Box 5, Folder 80, SHC. Also "The Cane Growers' Association of the United States," *LP*, XIX (1897), 90–91. D. D. Colcock to Mattie (wife), April 24, 1896, in Folder 36, Daniel De Saussure Colcock Papers, Manuscripts Department, Special Collections Division, Howard-Tilton Library, Tulane University.

11. "The Louisiana Sugar Planters' Association," *LP*, XVIII (1897), 163–64.

the Republican party. The cane growers' chief lobbyist in Washington, D. D. Colcock, worked hard during the spring of 1897 to raise the sugar duties, eventually succeeding with the Dingly tariff.[12]

Although a concerted effort to increase the LSPA membership between 1897 and 1900 was decidedly a success, the organization's ambivalence about involvement in political affairs permanently weakened its effectiveness. The events of 1894 and 1896 had shown planters that profitable sugar manufacturing was a consequence of successful politics as well as the result of the introduction of scientific methods and advanced technology. Although Stubbs remained an active LSPA member, the instability of the sugar experiment station's chief patron caused him to depend on the state as a primary source of funds.

The Changing Nature of the International Sugar Market

Political dissension within the LSPA had splintered the Louisiana planter community. In addition to internal problems, the Louisiana sugar industry was also confronted with new economic challenges, the result of a rapidly changing international sugar trade.[13] Between 1865 and 1898, Louisiana and all other American sugar-producing areas had benefited from either a tariff or bounty. In 1876 Congress had extended the protective sugar duty to include Hawaiian sugar manufacturers. As a consequence, Hawaiian sugar exports to the United States increased markedly. The 1898 Spanish-American War ultimately led to the expansion of this tariff-free area to include not only Hawaii, but also Puerto Rico and the Philippine Islands. In addition, Cuba, which was closely tied both politically and economically to the United States, was granted a 20 percent reduction in the tariff rate in 1902. These concessions rapidly stimulated sugar production in the so-called producer islands and in part resulted in the replacement of European beet-sugar imports to the United States

12. D. D. Colcock to Charley [Farwell], February 21, 1897, Colcock to Mattie [wife], March 3, 1897, Colcock to Charles A. Farwell, March 7, 1897, Colcock to Nelson Dingley, March 6, 1897, all in Box 4, Folder 37, Colcock Papers, Tulane.

13. John E. Dalton, *Sugar: A Case Study of Government Control* (New York, 1937), 19–39; Lippert S. Ellis, *The Tariff on Sugar* (Freeport, Ill. [1933]), 23–72.

by cane sugar originating primarily from Cuba. As a result, Louisiana planters, who had modernized their industry in response to the competitive threat of the European beet-sugar industry, were faced with the emergence of new rivals. Cane sugar interests benefited not only from United States tariff agreements, but also from decisions made by European representatives to the 1902 Brussels Convention. Throughout the last half of the nineteenth century, the percentage of beet sugar in the international sugar market had increased steadily from 14.3 percent in 1850 to 65.5 percent in 1900. However, this trend was reversed on March 5, 1902, when representatives from Germany, Austria, Belgium, Spain, France, Great Britain, Italy, the Netherlands, Sweden, and Norway agreed to abolish all indirect and direct bounties on the production and exportation of beet sugar.

As a result, two production areas dominated the world sugar market during the first quarter of the twentieth century. Sugar grown and manufactured in Java was exported north and west to India and western Europe, while Cuban sugar was transported north and east to the United States and Europe. The secondary producers, Hawaii and the Philippines, supplied the west coast of the United States, and Puerto Rico supplemented Cuba in supplying raw sugar to large United States refineries in the East.

Within this international framework, most of the refined Louisiana sugar was shipped up the Mississippi River to major midwestern markets in St. Louis, Memphis, and St. Paul. The rise of a new international sugar trade was only one factor behind the serious depression experienced by the Louisiana sugar industry between 1906 and the beginning of World War I. Falling prices, disastrous crop freezes, heavy floods, serious labor shortages, and the spread of plant diseases all contributed to a marked decline in production. These conditions led the usually optimistic Stubbs to describe the once prosperous Louisiana sugar bowl as "Tis Greece, but living Greece no more." For most planters, sugar manufacturing was no longer a profitable occupation. Sugar planter and former United States senator Donelson Caffery complained in 1904, "If I could only get free from the sugar business, without losing everything, I would bless my good stars—as it is, I am bound to keep the plantation up, hoping for a chance to sell." By 1906 Caffery was convinced that he should establish a cattle ranch or pecan plantation, and he encouraged his sons to prospect for oil.[14]

During this period of turmoil, the once-powerful LSPA was unable to advance Louisiana interests in Washington, to campaign for a strong research program at the Louisiana Sugar Experiment Station, or to protect planters against the unfair practices of the American Sugar Refining Company.[15] After 1900 approximately 80 percent of all raw sugar produced in Louisiana was sold through the Louisiana Sugar Exchange to this trust based in New York City. Between 1900 and 1910 the American Sugar Refining Company's refinery in Chalmette, Louisiana, purchased plantation-grade sugar at prices below the New York market price. Using both threats and secret agreements, the American Sugar Company's leaders discouraged other New York refiners from competing for Louisiana's raw sugar. The LSPA did little to oppose the trust's activities. For example, Louisiana planters could have organized their own cooperative refinery. Instead, they accepted whatever price the trust offered to avoid the possibility of a boycott of their goods. Thus, the LSPA had not only surrendered its function as lobbyist for the industry, but also had lost the control it once had over the sugar market.

The Louisiana Sugar Experiment Station, 1905–1912

The decline of the LSPA had marked consequences for research work at the sugar experiment station. The LSPA failed to fund the station after 1905, but federal money provided by the Adams Act guaranteed that research in sugar manufacturing would continue at Audubon Park. In November of 1904, W. R. Dodson, a Harvard-educated botanist, was appointed as Stubbs's successor; and R. E. Blouin, assistant director of the Louisiana Sugar Experiment Station under Stubbs's

14. William C. Stubbs, Undated Address (Typescript in Stubbs Papers, Folder 124, College of William and Mary); Donelson Caffery to daughter, January 10, 1904, in Donelson Caffery and Family Papers, Folder 10, SHC; Caffery to daughter, September 2, 1906, in Caffery Papers, Folder 11, SHC.

15. For information on Louisiana and the Sugar Trust, see U.S. Congress, House of Representatives, *Hearings held Before the Special Committee on Investigation of the American Sugar Refining Co. and Others* (Washington, D.C., 1911), II, 1757–1910.

leadership, was given responsibility for the day-to-day activities at the New Orleans installation. Under Dodson's leadership, the focal point of state experiment station work gradually shifted from New Orleans to Baton Rouge. Fertilizer and soil analysis, once an important activity at Audubon Park, was conducted at the Baton Rouge station laboratory after 1905. Consequently, the several assistant chemists assigned primarily to routine work, but who periodically assisted in research investigations at the sugar experiment station, were transferred to Baton Rouge. In 1903 the sugar experiment station employed twelve professional employees; but by 1906 only six were assigned to the staff. Within the span of a few years, the station was transformed from a center of scientific activity to a second-class institution.

Between 1906 and 1908, while the LSPA still remained a somewhat active organization, a close working relationship existed between scientists and planters. The station's research chemist, Fritz Zerban, who had received his doctorate at Göttingen, conducted valuable studies on the determination of sulfur compounds in sugar products and investigated clarification processes.[16] However, after 1908 the bond between scientists and planters appears to have been broken. And this break seems to have diminished the fortunes of both parties. Planters needed scientific expertise more than ever if they were to revive the declining industry.

In 1908 rumors circulated that the unimportant Audubon Park sugar experiment station would soon move all operations to Baton Rouge. Apparently, Dodson favored the abolishment of the Audubon Park station, for he allocated insufficient funds to maintain its physical plant.[17] In a short time, the ornamental gardens surrounding the station were choked with weeds, and the once-imposing buildings were in dire need of cleaning, painting, and repairs. The sugarhouse had deteriorated so badly that Dodson remarked in the 1908 report to the governor that it "is now in dire need of extensive

16. Fritz Zerban, "Sulphur in the Process of Sugar Manufacture," *Sugar Planters' Journal*, XXXVII (1907), 304; Fritz Zerban, "Clarification of Cane Juices in Louisiana," *LP*, XXXVIII (1907), 171; Fritz Zerban, "Preliminary Investigation on the Composition of a Cuban Raw Sugar," *LP*, XXXIX (1908), 13; Fritz Zerban, "Studies on the Sulphur Control in the Sugar House," *LP*, XXXIX (1908), 388.

17. R. E. Blouin to C. A. Browne, May 12, 1908, in Browne Papers, Box 2, LC.

repairs, and the installation of modern machinery. . . . Extensive improvements will be necessary if the station is to continue its sugar house investigations."[18]

While pursuing an extensive research program, Stubbs had proved himself an able manager of both funds and personnel. Unfortunately, Dodson was incapable of dealing with university administrators, politicians, and even his own subordinates. Between 1905 and 1912 both the station director and the station itself were the subject of many complaints. It was said that the station had "too many bosses"; it lacked "strong, efficient control" and was directed by an "unsympathetic and unappreciative administration." Further, it was remarked that Dodson could not "handle money or men." This apparent lack of managerial ability and control quickly interfered with research. For example, in 1905 C. A. Browne was assigned a problem in rice nutrition, but the project was almost impossible to complete. Since the necessary laboratory facilities and support personnel were no longer in New Orleans, Browne would be required to travel to Baton Rouge, start the experiment, and then entrust the work in progress to a station scientist unfamiliar with the details of investigation.[19]

Perhaps as a consequence of the confusion surrounding authority at the sugar experiment station, the turnover rate among professional personnel increased to the point that it seriously affected the quality of research conducted there. Station chemists appeared to have difficulties in problem definition, and an incoherent, irrelevant research program emerged. Chemists generally pursued research that was of personal interest, but the knowledge gained was frequently of little use to planters. P. A. Yoder, a Göttingen graduate and station chemist from 1908 to 1910, never focused his efforts on any one major problem; he eventually published a paper on the occurrence of formaldehyde in sugarhouse products that was probably regarded as an insignificant contribution by both planters and sugar chemists. W. E. Cross, also a Göttingen graduate, served at the station from 1911 to 1914. While his analytical work was first-rate in

18. W. R. Dodson, *22nd Annual Report of the Agricultural Experiment Stations of the Louisiana State University for 1909* (Baton Rouge, 1910), 8.

19. P. A. Yoder to C. A. Browne, June 17, 1910, in Box 19, Browne to Yoder, June 24, 1910, in Box 19, J. E. Halligan to Browne, in Box 8, all in Browne Papers, LC. Browne to C. F. Langworthy, November 29, 1905, in Louisiana File, RG 164, NA.

its objectives and execution, it was beyond the instrumental capabilities of the typical plantation sugarhouse laboratory.[20]

Even under Stubbs's guidance, chemists and assistant chemists frequently found higher paying, more responsible positions at other experiment stations or in private industry. However, after 1907 it appears that several employees left the station not for better positions, but to avoid an unpleasant work situation. During 1907–1908 four chemists left the sugar station, including the assistant director R. E. Blouin and chief chemist Fritz Zerban. Blouin was appointed director of the Argentina experiment station, while Zerban accepted a similar position in Peru. Assistant chemist J. A. Hall joined former chemist Charles Browne at the New York Trade Laboratory, while several other members of the station staff found employment with the Cuban-American Sugar Company. Even Dodson himself voiced dissatisfaction with his own job. He felt that President Boyd was neglecting the agricultural department at Louisiana State University in Baton Rouge, and in July of 1910 Dodson resigned. Boyd refused to accept Dodson's resignation, however, and promised him more administrative assistance. Yet, while Dodson had at least temporarily reached an understanding with Boyd, he had a violent disagreement with the commissioner of agriculture in 1912. Since Stubbs's retirement, the Board of Agriculture, which had legal accountability for the station, had failed to meet. The Louisiana State University Board of Supervisors had assumed the responsibility for making decisions on station policy, and apparently, until 1912, the commissioner of agriculture routinely signed the financial statements sent to the Office of Experiment Stations in Washington.[21] In 1912 the commissioner of agriculture refused to sign the experiment station's account sheets. However, Dodson temporarily circumvented the commis-

20. W. E. Cross to Browne, December 29, 1914, in Browne Papers, Box 3, LC; W. E. Cross, (Technical) Investigations on Methods of Analysis of Cane Products, Louisiana Experiment Station Bulletin CXXXV (Baton Rouge, 1912); W. E. Cross, Clarification of Louisiana Cane Juices, Louisiana Experiment Station Bulletin CXLIV (Baton Rouge, 1914).

21. Dodson to A. C. True, July 8, 1910, and July 26, 1910, in General Correspondence, File 50, Dodson to True, September 11, 1912, in State Correspondence, all in RG 164, NA. Dodson to True, June 11, 1912, in RG 164, State Correspondence, Administration and Policy Accounts, NA.

sioner's act by persuading the Louisiana traveling auditor of public accounts to certify station finances. The quarrel between Dodson and the commissioner probably stemmed from Dodson's efforts to reform the state's fertilizer inspection law. State inspectors were appointed by the commissioner, and in 1910 their total salaries were double that of the entire expenditures of the experiment station.[22] Before the argument ended, Dodson's authority was undermined by the defection of one of the station's chemists, who sided with the commissioner.[23]

Thus, the cooperative network that had existed between the LSPA and the Audubon Sugar Experiment Station from the 1890s to the early twentieth century, evidenced in the work of Stubbs, Beeson, Maxwell, and Browne, had disintegrated. Sugar chemistry conducted at the station after 1908 was characteristically disoriented and of little consequence to an industry desperately in need of scientific assistance for its survival.

Sugar Engineering and Chemical Engineering in Louisiana, 1904–1912

The sugar curriculum, originated at the LSPA-sponsored sugar school in 1891 and transferred to Louisiana State University in 1897, changed considerably between 1903 and 1908. Theoretical precepts directly applicable to processes in the sugarhouse were gradually injected into a program once characterized by practical training. Stubbs's university position as professor of agriculture had been at least partially dependent upon political and economic support from the steadily declining LSPA. After 1900 he found that his educational philosophy was under attack by university president Thomas Boyd, who stressed the value of theoretical rather than practical knowledge. Additional courses in carbohydrate chemistry and mechanical engineering were added at the expense of the agricultural studies. With Stubbs's retirement in 1905, the sugar course was changed to a program in sugar engineering.[24] As originally conceived by Dymond,

22. W. R. Dodson, *Twenty-Sixth Annual Report of the Agricultural Experiment Stations of the Louisiana State University for 1913* (Baton Rouge, 1914).

23. J. E. Halligan to Browne, May 22, 1913, Box 8, Browne to H. P. Agee, January 26, 1914, Box 1, all in Browne Papers, LC.

24. Louisiana State University, *Catalogue, 1905–06* (Baton Rouge, 1906), 29.

Stubbs, and others within the LSPA, the sugar expert had to be knowledgeable in the agriculture of sugarcane. By 1908, however, agriculture, which at one time had been an important component of the sugar course, had been given a secondary role in the curriculum. Stubbs, then retired, lamented the lack of interest by Louisiana State University administrators to support agricultural education. "Go into the university and look around, attend exercises, interview the students, and you will find little or no agricultural atmosphere. An agricultural college must have an atmosphere of its own. Highly polished shoes, fashionably fitting clothes and unsoiled hands must be substituted by rubber boots, overalls and working gloves."[25] Stubbs had maintained a nineteenth-century view of agricultural and mechanical education that placed an emphasis on the dignity of labor. However, university administrators were apparently modifying their educational views at the turn of the twentieth century, and farm work was supplanted by laboratory exercises in the curriculum.

Mechanical engineering practices and theory played an increasingly important part in training the sugar engineer. The fourth-year student was taught three general machine-design courses, and during his fifth year he specialized in the design of sugar equipment.[26] Peculiar to the training of the sugar engineer was an emphasis on understanding the evaporation process in detail, from both a theoretical and practical standpoint. At the Audubon Sugar School, the graduate thoroughly understood the structural and process design of the vacuum evaporator. This narrow focus in the student's training stemmed from the economic significance of the evaporation step in the manufacturing process. Next to the value of the cane, the cost of fuel was the most expensive element in sugar manufacturing. At the sugar school, the theoretical understanding of the process of evaporation had its basis in a treatise written by E. Hausbrand, entitled *Evaporating, Condensing and Cooling Apparatus.*[27]

25. William C. Stubbs, Undated Address [1915?] (Typescript in Stubbs Papers, SHC).

26. "Problems in Machine Design I" and "Problems in Steam Machinery IV" (Typescripts in Anatole J. Keller Papers, Folder 18, Special Collections, Hill Memorial Library, LSU); Keller, "Experimental Engineering" Notebooks (MS in Keller Papers, LSU).

27. E. Hausbrand, *Evaporating, Condensing and Cooling Apparatus*, trans. A. C. Wright (2nd Eng. ed., London, 1916).

Hausbrand's approach was rooted in nineteenth-century technical innovations closely associated with the sugar industry, and in particular, he drew upon the work of Paul Horsin-Dèon. Horsin-Dèon, Norbert Rillieux's personal secretary and business associate, published several treatises in which physical laws were applied to the understanding of the evaporation process, and these works were used as textbooks in advanced courses at both Louisiana State University and Tulane.[28] In his 1882 work, Horsin-Dèon began by describing the opposition that the theoretician normally received in the chemical industry, and he stated that his goal was to introduce parameters that would account for the nonideal conditions normally found in the manufacturing plant. In his discussion of evaporation, he calculated the theoretical heating surface of multiple-effect evaporators by using the heat equations of Regnault and Mason, along with equations of heat transfer. Horsin-Dèon's 1900 treatise dealt with the same subject, but on a higher level of mathematical sophistication. He described the process taking place inside the evaporator in terms of a dynamic equilibrium, one in which vapor from the syrup was both evaporating and condensing. Horsin-Dèon's contributions were in his use of mathematical analysis, kinetic theory, and the ideal gas laws in deriving empirical relationships. His work contained many of the precepts later refined in Hausbrand's treatise.

Hausbrand successfully applied the basic laws and principles of physical science to the solution of particular industrial problems. In the opinion of an eminent chemical engineer and historian, Hausbrand was the "father of modern chemical engineering treatment of unit operations."[29] He applied theories of heat transfer, the thermodynamic properties of steam, parallel and counterflow heat exchange, and the significance of logarithmic mean temperature differences to problems in process design.

Furthermore, E. W. Kerr, a nationally known mechanical engineer, employed Hausbrand's approach in his lectures and research publications.[30] Kerr was a graduate of the Texas Agricultural and Mechanical

28. Horsin-Dèon, *Traité Théorique et Pratique de la Fabrication du Sucre*, 499–557.

29. W. K. Lewis, "Evolution of Unit Operations," *Engineering Progress*, LV (1955), 3.

30. Kerr's publications include: *Preliminary Tests of Sugar House Machinery*, Louisiana Experiment Station Bulletin CVII (Baton Rouge, 1908); *An Experimental*

College, receiving a bachelor's degree in 1896 and his master's in 1899. He had also taken special engineering courses at Stevens Institute, the University of Wisconsin, and Purdue. As an assistant professor at Texas Agricultural and Mechanical College from 1896 to 1902, he wrote a popular treatise entitled *Power and Power Transmission*. Later he taught at Purdue, and in 1905 he joined the faculty at Louisiana State University as professor of mechanical engineering.[31]

With the frequent assistance of his students, Kerr conducted systematic research that provided the foundation for understanding the mechanical interrelationships between various components of the cane-sugar manufacturing process. Kerr's most notable article prior to 1914 was entitled *Performance Tests of Sugar House Heating and Evaporating Apparatus*. This study of the variables affecting heat transmission and entrainment resulted in the quantification of large-scale physiochemical phenomena as a function of design. Therefore, in response to a practical problem, Kerr and his student engineers accepted specific scientific laws and, in applying them, unified chemical and mechanical engineering knowledge.

While Kerr's studies united chemical and engineering knowledge, Charles Coates, professor of chemistry, generally failed to employ mechanical methods or to rely on the principles of mechanics in his work. He assumed that the design of apparatuses was fixed and that his role was to improve the chemical aspects of the process by utilizing data gained through analytical chemical methods. Process modification on the large scale was done either by altering existing reaction conditions or by proposing new reactions.

At the Audubon Sugar School in Baton Rouge, mechanical engi-

Study of Bagasse and Bagasse Furnaces, Louisiana Experiment Station Bulletin CXVII (Baton Rouge, 1909); *Bagasse Drying*, Louisiana Experiment Station Bulletin CXXVII (Baton Rouge, 1911); *Experiments with Oil Burning in Boiler Furnaces*, Louisiana Experiment Station Bulletin CXXXV (Baton Rouge, 1911); *An Experimental Study of Heat Transmission*, Louisiana Experiment Station Bulletin CXXXVIII (Baton Rouge, 1913); *Performance Tests of Sugar House Heating and Evaporating Apparatus*, Louisiana Experiment Station Bulletin CXLIX (Baton Rouge, 1914); and *Experimental Studies of Vacuum Juice Heaters*, Louisiana Experiment Station Bulletin CLIX (Baton Rouge, 1916).

31. "Statement of Training and Experience of E. W. Kerr" (Typescript in E. W. Kerr Papers, Special Collections, Hill Memorial Library, LSU).

neers assimilated the physiochemical knowledge necessary to understand the various large-scale processes. Successfully uniting chemistry and engineering, they consequently practiced chemical engineering. However, because the chemists at the Audubon school resisted employing mechanical principles to aid their research, they remained cast in the role of mere chemists.

At the turn of the century, the sugar course at Louisiana State University served the needs of the LSPA. However, with the 1901 discovery of oil at Spindletop, Texas, and the emergence of new organic extractive industries, the university instituted a chemical engineering program in 1908 that was modeled after the sugar engineering course. The school was now prepared to meet not only Louisiana's current requirement for sugar engineers, but also the perceived future demand for chemical engineers capable of working in industries other than sugar manufacturing.

Ironically, it appears that Louisiana State University's chemical engineering program had closer ties to sugar chemistry than sugar engineering. The 1908 catalog stated that the Audubon Sugar School "has for many years offered a course in the chemical engineering of cane sugar." The purpose of the new course was to be "more general in its nature . . . [to] train the student in those fundamental principles which underlie all engineering problems."[32] In contrast to the sugar school, where it was not required, physical chemistry was the focal point in the chemical engineering course. Theory, not practice, was emphasized. Indeed, the chemical engineering students were never given the opportunity to familiarize themselves with the operation of chemical machinery. Their practical experience consisted solely of observations during plant tours. By placing primary emphasis on all branches of chemistry and giving only slight exposure to a broad range of unrelated engineering courses (including electrical engineering, physics, and mechanical engineering), Louisiana State University effectively trained its chemical engineering students to be primarily chemists with a general engineering background.

Although the university's chemical engineering program did not attract many students, over one hundred pupils were enrolled in the sugar engineering program in 1912. Even though the Louisiana sugar

32. Louisiana State University, *Catalogue, 1908–09,* 76.

industry experienced a period of decline after 1906, the Audubon Sugar School continued to be popular among students. This trend was in part attributed to the emergence of Cuba as "the new sugar bowl of the world." After the Spanish-American War, American financial and commercial interests directed industrial development in Cuba. A 1916 popular book proclaimed: "Cuba has much modern sugar machinery and many modern mills; she needs all these, and more like them; but Cuba's big opportunities are not alone for the machinery salesman; the big opportunities she presents are for enterprise and efficiency." The author continued by stating that the combination of technical expertise and capital "shall lift her [Cuba's] chief industry out of its Spanish lethargy."[33] As an example of American influence in the Cuban sugar business, in Oriente Province during 1913–1914, fourteen of the thirty major plantations were American-owned, and they employed thirty-nine chemists.[34] Thus, it is no surprise that many graduates of the Louisiana State University Audubon Sugar School found employment in this rapidly expanding industry less than six hundred miles from Louisiana. Indeed, the change in the balance of agricultural and engineering courses in the Audubon Sugar School's curriculum may have been at least partially in response to the demands of the Cuban sugar industry. The challenges facing the Cuban sugar interests were not so much in agriculture—planting conditions there were ideal—but in the design of large, high-volume central factories.

Although Tulane University had no formal connections with the LSPA, its sugar engineering curriculum experienced changes similar to those at the Audubon Sugar School at Louisiana State University. While Tulane did not espouse the educational philosophy associated with nineteenth-century agricultural and mechanical schools, this university had based its nineteenth-century engineering education program on manual training. As in the case of agricultural farm labor at the state university, shopwork was replaced in the Tulane curriculum by more theoretical subjects after 1900. In addition, after 1910 the relative proportions of chemical and mechanical subjects changed

33. Robert Wiles, *Cuban Cane Sugar* (Indianapolis, 1916), 82, 83.

34. See Publica de Cuba, Secretaria de Agricultura, *Comercia Y Trabajo, Industria Azucarera de Cuba, 1913–1914* (N.p., n.d.).

due to a new emphasis on instruction in physical chemistry. This shift in Tulane University's educational philosophy was in part due to the efforts of its president, E. B. Craighead. In his 1905 inaugural address, Craighead stated: "Louisiana holds the wealth of a world within her grasp. . . . But how may Louisiana grasp and improve her opportunity? The answer is immediate. By giving to her youth the best and most practical industrial, as well as the highest scientific training. . . . But to render practical training more and more efficient, there must be encouraged and fostered the high scientific training of the university."[35]

By 1906, shopwork, which had been the key to the early training of engineers at Tulane, had become almost nonexistent in the new curriculum. Instead, chemical and sugar engineers were required to take an increasing number of physics and physical chemistry courses. Courses in mechanics and heat, electricity, sound and light, electricity and magnetism, and thermodynamics were required of all engineering students. The chemical engineering program differed from that of sugar engineering by the substitution of physical chemistry, electrochemistry, and chemical plant design in lieu of sugar machinery design, sugar house erection, sugar house layout, and power house design. Thus, chemical engineering at Tulane had changed from a course dominated by engineers to one that focused on the physical sciences.

Perhaps to divide practical from theoretical instruction, a special two-year course was instituted for those students who were previously considered special students in the program. Nonmathematical courses in general chemistry, qualitative and quantitative analysis, technical analyses, agricultural chemistry, agricultural analysis, and the steam plant and sugarhouse were offered. The new program enabled Professor Levi Wilkinson to teach the fundamental principles of chemical analysis and agricultural chemistry to a group of pupils generally less prepared to learn advanced mathematics.

Wilkinson resigned from Tulane in 1912. Although his successor, A. O. Williamson, continued to teach a course in industrial chemistry, he also failed to unite chemistry with engineering knowl-

35. Tulane University, "Installation of President Edwin Boone Craighead, March 16, 1905," *Tulane University Bulletin* (1905), 36–37.

edge. Practice, which once served to join these two disciplines, was minimized in this Tulane program, and it remained for engineering courses in evaporative machinery design and evaporative machinery to draw upon principles that would later be considered to be peculiar to chemical engineering. While Tulane engineering faculty recognized that practical training was a necessary component of engineering education, they felt that such knowledge should be acquired in the engineering laboratory, rather than in the shop. William B. Gregory directed the experimental engineering courses at Tulane, and at least on one occasion he made overtures to the LSPA to secure funding for an evaluation of sugarhouse fuel consumption.[36] Gregory was attempting to extend the methods of experimental mechanical engineering from merely measurements of shaft speeds and physical tolerances to evaluations of sugarhouse apparatuses.

The Tulane University program in sugar and chemical engineering also expanded as a result of American industrial efforts to develop the sugar industry in Cuba. After 1908 an increasing percentage of the Tulane students came from Cuba; between 1909 and 1914 about 25 percent of the pupils in sugar engineering were Cuban.

The sugar engineering and chemical engineering programs at Tulane and Louisiana State University developed somewhat comparably. During the 1890s both schools emphasized the practical aspects of technical education. At Tulane, the Manual Training Program, with its emphasis on shop skills, served as the basis for training engineering students. Thus, Tulane's initial chemical engineering program fused a foundry tradition with theoretical scientific and technical knowledge. At the Audubon Sugar School, first located at the New Orleans Sugar Experiment Station, students united chemical and engineering knowledge while working in the sugarhouse. In the classroom at both institutions, engineers rather than chemists applied physical laws to the solution of practical manufacturing problems. Yet, by 1910 chemical and sugar engineering at Tulane and Louisiana State University had become less practical and more theoretical. Shopwork and agricultural subjects were replaced by courses in the physical sciences, and particularly by physical chemistry and electrochemistry. Indeed, at least one reason why both universities

36. "The Louisiana Sugar Planters' Association," *LP*, XLII (1909), 97.

shifted to more theoretical engineering programs was that with the decline of the LSPA an institutional patron of the practical arts no longer existed, and one was needed. While the LSPA played a crucial role in fostering engineering education in Louisiana during the early 1890s, the organization's influence was short-lived. Controversial national party policies ultimately led to a breakdown in the traditional political consensus within the LSPA, and the association subsequently fragmented. Toward the close of the first decade of the twentieth century, the remaining LSPA members became increasingly pessimistic over the future of the Louisiana sugar industry. The enthusiasm of the monthly meetings held during the 1890s was notably absent. The industry was contracting rather than expanding, becoming a source of opportunity for only a handful of entrepreneurs.

In January, 1909, the monthly meeting featured a ceremonial presentation of oil portraits of the LSPA's past presidents. The speeches and reminiscences marked the end of one generation's contributions to industrial growth in Louisiana. One of the most poignant was by Henry McCall.

> Those of us who have retired from strenuous activity of the planting life, will be content to turn over the reins of government to younger men. Some time since, I was deploring the fact that neither of my two boys were taking up the work at Evan Hall, which four generations had accomplished. One of my boys answered, "You quit, Papa." "Yes," said I, "when you have devoted 35 years of your best life to that dear old plantation, you can quit too."[37]

Those LSPA founders still alive, like Henry McCall and John Dymond, remembered deceased former leaders Kenner, Brent, and others. The older generation toasted their own achievements and symbolically passed the yoke of future responsibilities to the group of younger leaders. Unfortunately, their successors were unable to surmount difficulties no less imposing than those that had accompanied Reconstruction. The LSPA then became more of a social club than an institution for the exchange of scientific and technical ideas. Meetings were often canceled because of a failure to gather a quorum,

---

37. "Louisiana Sugar Planters' Association: Presentations to the Association of the Portraits of Etienne de Bore and of the Past Presidents of the Association," *LP*, XLII (1909), 59–62.

and in 1910 all but two meetings were no more than social events. When meetings were held, the speakers invariably came from the state experiment station.[38] Louisiana sugar planters in attendance apparently lacked the initiative and enthusiasm that had been so evident at the LSPA meetings of the 1880s and 1890s. In 1917 membership fees in the LSPA were dropped, yet even this move failed to add members to the rolls. The organization held only one meeting between 1917 and 1920, and in August of 1922 the LSPA merged with the American Cane Growers' Association to form the American Cane Sugar League. Although its leading figures were dead, including Dymond and Stubbs, the LSPA continued to exist as a legal corporation until May 6, 1931, when thirteen members solemnly met and liquidated its assets. Despite its relatively short lifetime, the LSPA had left an indelible mark on Louisiana.

Of all LSPA contributions made to the Louisiana sugar industry during the late nineteenth and early twentieth centuries, the organization's efforts in the development of engineering education was the most enduring. The peculiar combination of agricultural, chemical, and engineering knowledge associated with the independent Audubon Sugar School was later assimilated and modified by Louisiana State University, which had broader interests and objectives.

In a very real sense, the Louisiana Sugar Planters' Association initiated sustained technological change in Louisiana. Between 1885 and 1895 a scientific and technical infrastructure had emerged in that state, and this network of scientific institutions possessed an inherent flexibility not evidenced in the LSPA. As the economy of the state diversified, the LSPA-sponsored organizations fell under state control and assumed different strategies and objectives to meet new economic opportunities. As a result, this change in institutional setting led to the development of a broad-based theoretical program in chemical engineering to produce scientific professionals whose skills could benefit a variety of industries.

38. See Louisiana Sugar Planters' Association, "Minutes of Monthly Meetings, 1891–1931" (MS in Southwestern Archives and Manuscripts Collection, USL), 12–51.

12

The Anatomy of an Industry in Transition

To live in the modern world means to cope with the complexities of change. Similar to our situation today, late-nineteenth-century Louisiana sugar planters John Dymond, Duncan Kenner, and others were caught in a world of changes, some visible and some invisible, and they were trying to make sense of it the best way they could.[1] Rather than succumbing to the impersonal forces of their environment, LSPA leaders responded to a threatening economic situation by seeking to control their surroundings.[2] In so doing, the LSPA proved to be a major stimulus to the modernization of the sugar industry.

The indiscriminate use of the term *modernization* frequently makes scholars flinch.[3] In this study, modernization can be equated with the response of the Louisiana planters to competition. After these men organized, they began to recognize that science had power. Through their efforts, the Louisiana sugar industry's business practices and production methods were brought up to date in order that they might compare favorably with those in Europe and other important cane-producing centers of the world.

1. Robert Wiebe, *The Search for Order, 1877–1920* (New York, 1967), 11–43.
2. Galambos, "Technology, Political Economy and Professionalization," 492–93; Howard E. Aldrich, *Organization and Environments* (Englewood Cliff, N.J., 1979), 323–50.
3. *International Encyclopedia of the Social Sciences*, X, 386–95.

The planters who banded together in 1877 to establish the LSPA viewed organization as a means by which stability and economic survival could be achieved. In its early years the LSPA was held together by many factors, including kinship ties, neighbor relations, social club membership, and mutual economic interests. Two institutional objectives emerged: a desire to control an extremely complex economic environment and a drive to promote science-based technology related to the sugar industry.

Immediately after forming the organization, LSPA leaders felt that political means would be the best way to gain control of their environment. Through LSPA lobbying efforts in Washington and by securing a position for one of its members on the United States Tariff Commission, satisfactory tariff levels on the grades of sugar produced in Louisiana were maintained during the 1880s, and a profitable sugar bounty was enacted between 1890 and 1894. These federal laws resulted in short-term stability for sugar planters, thus giving them sufficient confidence to expand acreage under cultivation and to invest in new machinery. However, the most astute planters, including John Dymond, recognized that politics and congressional influence were of limited and perhaps transitory value in correcting either long-term threats from the European beet-sugar trade, or demands for lower prices within the United States brought about by the transformation of the workingman's diet. Many plantation owners recognized that shrewd politics should be supplemented by the application of science and science-based technology. Only by joining these two strategies could the Louisiana sugar industry be saved.

Planter response in placing high hopes on science was rather remarkable for the times. The idea that scientific knowledge could be applied to the improvement of manufacturing efficiencies and scale was in its infancy. Indeed, during the last quarter of the nineteenth century the place of the scientist within industry was only gradually evolving. Throughout the period, American business generally relied on the efforts of the self-educated inventors like Thomas Edison rather than university-trained experts. The latter were often viewed by the public as being impractical when confronted with the problems of the real world. Organized industrial research, whether at DuPont or at General Electric became significant only during the decade preceding World War I.[4] The science-based technology and research that took place during the modernization of the Louisiana sugar industry anticipated twentieth-century developments that would occur after the rise of big business.

Traditionally, the sugar planters associated politics with control. After 1880, however, science was also seen as a means of control. But the problem facing LSPA leaders was how to acquire the scientific and technical expertise necessary to gain this desired goal. During the late 1870s and early 1880s, the LSPA shifted its focus from politics concerning the tariff to politics involving science, and after an initial disappointment with local institutions, the LSPA engaged the United States Department of Agriculture in Louisiana affairs.

The relationship between the USDA and LSPA during the 1880s proved to be fruitful for both organizations. Chief chemist Harvey Wiley used LSPA support to sustain his position in Washington at a time in which he was coming under increasing attacks by a growing number of critics. And the Bureau of Chemistry's emerging program in the analysis of adulterants in foods was at least in part the consequence of Wiley's inability to master the complex problems related to the large-scale manufacture of sugar. Conversely, the USDA influenced LSPA policies by urging the association to establish a private sugar experiment station and later by providing this facility with financial assistance.

This drive to innovate led to the creation of an experiment station that for a short time—1890 to 1895—became one of the international focal points for research in carbohydrate chemistry. Station chemists trained at Harvard, Johns Hopkins, Zurich, and Göttingen explored the nature of organic reactions taking place both within the sugarcane plant and also in the large-scale processes of the sugar-

4. Leonard S. Reich, "Research, Patents and the Struggle to Control Radio: A Study of Big Business and the Uses of Industrial Research," *Business History Review*, LI (1977), 208–35; Leonard S. Reich, "Industrial Research and the Pursuit of Corporate Security: The Early Years of Bell Labs," *ibid.*, LIV (1980), 504–29; Leonard S. Reich, "Irving Langmuir and the Pursuit of Science and Technology in the Corporate Environment," *Technology and Culture*, XXIV (1983), 199–221. See also John W. Servos, "The Industrial Relations of Science: Chemical Engineering at MIT, 1900–1939," *Isis*, LXXI (1980), 531–49.

house. Concurrently, the LSPA, under the leadership of John Dymond, established a sugar school on the grounds of the experiment station with the hope that sons of planters would enroll and study chemistry and engineering. Science was power, and it was hoped that this power could be kept within the bounds of the planter class. It was at the Audubon Sugar School that everyday problems associated with manufacturing were studied, and as a result of this unique educational environment a new hybrid knowledge combining chemistry and mechanical engineering gradually emerged. The demands of a rapidly modernizing industry led not only to the call for a new kind of professional, but also a new discipline, sugar engineering, which later became a specialized branch of chemical engineering.[5] This linking of an industry in transition to the creation of a new profession and discipline was not confined to the late-nineteenth-century bayous of Louisiana. At the same time that Harvey Wiley and his USDA chemists were trying to combine chemistry and mechanical engineering at Warmoth's Magnolia plantation, scientists like George Davis were experiencing the same challenges within the context of another modernizing chemical industry, the British alkali trade.[6]

The alkali industry—manufacturers of such basic chemicals as sodium carbonate, sodium hydroxide, chlorine, and hydrochloric acid—was central to the economy of Victorian Britain. From its introduction in Britain in 1823 to the mid-1880s, the Leblanc method was the primary process of the trade.[7] Originating in France in the

5. Histories of chemical engineering include: Terry S. Reynolds, *75 Years of Progress: A History of the American Institute of Chemical Engineers, 1908–1983* (New York, 1983); William F. Furter (ed.), *A Century of Chemical Engineering* (New York, 1982); William F. Furter (ed.), *History of Chemical Engineering* (Washington, 1980); F. J. Van Antwerpen and Sylvia Fourdrinier, *Highlights of the First Fifty Years of the American Insitute of Chemical Engineers* (N.p., 1958); Alfred H. White, "Chemical Engineering Education in the United States," *Transactions of the American Institute of Chemical Engineers,* XXI (1928), 55–85. See also John A. Heitmann, "Two Examples of Early Chemical Engineering Education in America: The Audubon Sugar School of Louisiana and the University of Wisconsin" (paper delivered at the Society of the History of Technology Annual Meeting, 1980).

6. John A. Heitmann, "The British Alkali Industry, 1860–1890, and the Emergence of Chemical Engineering" (Paper, Johns Hopkins University, 1979).

7. C. T. Kingzett, *The History, Productions and Processes of the Alkali Trade*

late eighteenth century, the Leblanc process involved several distinct steps in which salt and sulfuric acid were converted to sodium carbonate, with the formation of hydrochloric acid as a by-product. During the third quarter of the nineteenth century, engineering innovations were applied and new processes developed so that the emitted hydrochloric acid could be trapped and converted to chlorine, which could be used as a bleach in the textile and paper industries. In the mid-1870s the improved Leblanc alkali industry remained the cornerstone of the chemical industry in Britain, but it was soon challenged by the new, more efficient Solvay process in a way not too dissimilar from the challenge of the European beet-sugar trade to cane growers in Louisiana.

In response to the Solvay challenge, the Leblanc soda makers were instrumental in forming the Society of Chemical Industry in 1882, and throughout the 1880s the old Leblanc factories modernized. The factory owner was especially concerned with lowering his fuel costs. Consequently, more efficient boilers, heat exchangers, and furnaces were designed. Waste products, like hydrochloric acid and calcium sulfide, were converted to marketable materials. Mechanical equipment—the conveyer belt, the automatic skimmer, the rotary filter, and the revolving furnace—were often included in new plant designs. Perhaps the best way to characterize the transition in the alkali industry between 1860 and 1890 from a technical perspective is to consider it as the concentrated application of machinery to a once-manual industry, the replacement of unskilled labor with machines. Like the Louisiana sugarhouse, the British Leblanc factory was filled with newly patented equipment.

Within this environment of change, and similar to developments in Louisiana, a chemist by the name of George Davis began to teach a new kind of knowledge that we now recognize as an early form of chemical engineering.[8] And just as sugar engineering emerged outside the regular educational framework in south Louisiana, so too in

---

(London, 1877); John Lomas, *A Manual of the Alkali Trade* (London, 1880); Kenneth Warren, *Chemical Foundations: The Alkali Industry in Britain to 1926* (Oxford, 1980).

8. On Davis, see "Obituary of George E. Davis," *Journal of the Society of Chemical Industry*, XXVI (1907), 598.

Britain Davis taught this new discipline during the 1880s at the Manchester Technical Institute and from his own consulting laboratory.[9] Like the students of the early Audubon Sugar School, Davis' pupils were generally outside the mainstream of higher education and were considered more as engineers than as chemists. Above all else these professionals were considered practical men.

Stubbs in New Orleans and Davis in Manchester both considered practical experience to be crucial to the solution of problems occurring in the large-scale factory. Stubbs wanted his students to get their hands dirty, while Davis asserted that "no matter what can be said in favor of calculations from purely theoretical data, prior experience will always be a favorite guide, and when we consider the mechanical conditions necessary for carrying out of chemical action on the large scale this experience carries our thoughts back to some process of manufacture with which we have been familiar."[10]

There are many parallels between the chemical engineering that emerged as the result of the transformation of the British alkali industry and a similar kind of applied science that developed in the south Louisiana sugar industry. The chief point, however, is that chemical engineering was the product of these two industries in transition, indeed, industries that modernized but nevertheless became noncompetitive by the early twentieth century. While the industries gradually died, the knowledge that was born during their competitive responses to external pressures found application in other areas of chemical technology.

While the focus of this study has been on the positive reception of science and science-based technology by the Louisiana Sugar Planters' Association, a good question to ask at this juncture is why did the Louisiana sugar industry fail to emerge as a mecca for research in the twentieth century, and why did the industry decline? Indeed, a crucial phase in modernization is the development of a self-sustaining economic system. In part the answer lies with the fragmentation of the LSPA because of internal political squabbles. However, two other factors also led to its eventual decline. First, the

9. His lectures were later published as George E. Davis, *A Handbook of Chemical Engineering* (2nd ed.; 2 vols.; Manchester, Eng., 1904).

10. *Ibid.*, I, 29.

industry was located in an inherently marginal cane-growing area, in a semitropical rather than tropical climate. Second, it was dependent on relatively expensive labor compared to such places as Cuba or Java. Like many other economic activities, sugar manufacture had its time in the sun, sustained during the late nineteenth century by politics on one hand and science on the other. Just like human beings, industries have their own lifespans, and for a brief time the Louisiana sugar industry rode a crest of induced prosperity. And similar to that of a successful man, the sugar industry left several legacies, including a new kind of applied science. This product of the trade at its zenith would later be modified to suit the needs of other extractive industries located in Louisiana.

Bibliography

Primary Sources

MANUSCRIPTS

Historic New Orleans Collection, New Orleans, La.
    Dymond Family Papers.
Library of Congress, Washington, D.C.
    Browne, Charles Albert. Papers.
    Wiley, Harvey W. Papers.
Louisiana State Museum, New Orleans
    Brent, Joseph L. Papers.
    Levert Family Papers.
    Riddell, William Pitt. "Riddell Genealogy." Record Group 10.
Special Collections, Hill Memorial Library, Louisiana State University, Baton Rouge
    Fleming, Walter L. Papers.
    Gay, Edward J., and Family. Papers.
    Keller, Anatole J. Papers.
    Keller, Arthur G. Papers.
    Kenner, Duncan. Papers.
    Kerr, E. W. Papers.
    Louisiana Sugar Planters' Association. "Minutes of Monthly Meetings, 1877–1891."
    McCall, Henry. "History of Evan Hall Plantation." Typescript, 1924.
    Pharr, John N., and Family. Papers.

Roy, Victor L. Papers.

Stewart, C. W. "A Brief Discussion of the Audubon Sugar Factory." Typescript, 1924.

Stubbs, William Carter. Papers.

National Archives, Washington, D.C.

Records of the Bureau of Agricultural and Industrial Chemistry. Record Group 97.

Records of the Commissioner of Agriculture. Record Group 16.

Records of the Office of Experiment Stations. Record Group 164.

Records of the Patent Office. Record Group 241.

Howard-Tilton Library, Tulane University, New Orleans, La.

Manuscripts Department, Special Collections Division

Barrow, Robert R., and Family. Papers.

Boston Club Collection.

Colcock, Daniel De Laussure. Papers.

Ferry, Alexis. Journal, Volumes 4 and 5.

Glenck, Robert. Papers.

Ordway, John Morse. Papers.

University Archives, Special Collections Division

Audubon Sugar School. Folder.

Ayres, Brown. Folder.

Ordway, John Morse. Folder.

Sugar Engineering. Folder.

Wilkinson, Levi W. Folder.

Louisiana Collection

Louisiana Sugar Industry. Scrapbook.

Southern Historical Collection of the University of North Carolina Library, Chapel Hill

Caffrey, Donelson, and Family. Papers.

Miles, William Porcher. Papers.

Shaffer, William A. Papers.

Stubbs, William Carter. Papers.

Warmoth, Henry Clay. Papers. Microfilm.

White, Maunsel. Papers.

Southwestern Archives and Manuscripts Collection, University of Southwestern Louisiana, Lafayette

Griffen, Lucile Mouton. Collection.

Louisiana Sugar Planters Association. "Minutes of Monthly Meetings, 1891–1931."

Swem Library, College of William and Mary, Williamsburg, Va.

Stubbs, William Carter. Papers.

CONTEMPORARY NEWSPAPERS

New Iberia *Louisiana Sugar Bowl*, 1870–1883.
New Orleans *Daily Picayune*, 1837–1880.
New Orleans *Times-Democrat*, 1881–1899.
New York *Daily Sugar Report*, 1882.

CONTEMPORARY PERIODICALS

American Chemical Journal, 1889–1893.
American Journal of Science, 1861–1865.
Annales de Chimie et de Physique, 1840, 1849.
Berichte der Deutschen Chemischen Gesellschaft, 1895.
Bulletin of the University of Wisconsin, Engineering Series, 1900.
Comptes Rendus de l'Academe des Science, 1846, 1850.
DeBow's Review, 1846–1864.
Deutsche Zuckerindustrie, 1880–1895.
Dingler's Polytechnisches Journal, 1870–1882.
Facts About Sugar, 1932.
International Sugar Journal, 1899–1915.
Journal of the American Chemical Society, 1879–1916.
Journal of Analytical Chemistry, 1887–1892.
Journal de Chimie Medicale, 1836.
Journal de Pharmacie et de Chimie, 1840, 1857.
Landswirtschaftlichen Versuchs-Stationen, 1891–1912.
Louisiana Planter and Sugar Manufacturer, 1888–1929.
Proceedings of the American Chemical Society, 1879.
Proceedings of the Association of Official Agricultural Chemists, 1891.
Proceedings of the Society for the Promotion of Engineering Education, 1898.
Sugar Cane, 1870–1898.
Sugar Planters' Journal, 1895–1910.

ANNUAL REPORTS, CROP STATEMENTS, AND
UNIVERSITY CATALOGS

Annual Report of the Agricultural Experiment Station of the Louisiana State University and A. & M. College . . . To the Governor. 1888–1911.
Bouchereau, A. *Statement of the Sugar and Rice Crops Made in Louisiana. . . .* 1878–1917.
Bouchereau, L. *Statement of the Sugar and Rice Crops Made in Louisiana. . . .* 1868–77.
Bright, Louis J. *New Orleans Price Current: Yearly Report of the Sugar and Rice Crops of Louisiana.* 1876–78.

Champomier, P. A. *Statement of the Sugar Crop Made in Louisiana in 1859–1860.*

Louisiana State University and Agricultural and Mechanical College. *Catalogue . . . .* 1899–1915.

State University of Louisiana. *Eighth Annual Session of the Collegiate Department: 1858–59.* New Orleans, 1859.

Tulane University. *Catalogue of the Academical Department.* New Orleans, 1885.

Tulane University, College of Technology. *Course in Sugar Engineering and Sugar Chemistry: Circular of Information.* New Orleans, 1899.

Tulane University. *Register. . . .* 1890–1917.

U.S. Department of Agriculture. *Report of the Commissioner of Agriculture. . . .* 1877–1892.

University of Louisiana. *Annual Report, University of Louisiana, February 14, 1861.* N.p., 1861.

University of Louisiana. *A Catalogue of the Offices and Students of the University of Louisiana for the Academical Year 1850–51.* New Orleans, 1851.

PAMPHLETS AND BOOKS

Agriculturists' and Mechanics' Association of Louisiana. *Proceedings of the Agriculturists' and Mechanics' Association of Louisiana: Annual State Fair, 5 January 1846.* New Orleans, 1846.

Aime, Valcour. *Plantation Diary of the Late Valcour Aime.* New Orleans, 1878.

Ames, H. O. *H. O. Ames' Improved Method of Evaporating Saccharine Juices by Steam.* New Orleans, 1859.

Alderman, Edwin Anderson. *Inaugural Address, March 12, 1901.* Boston, 1904.

Bache, A. D., and R. S. McCulloch. *Reports for the Secretary of Treasury of Scientific Investigations in Relation to Sugar and Hydrometers.* Washington, D.C., 1848.

Basset, Nicholas. *Guide Pratique De Fabricant De Sucre.* 3 vols. Paris, 1872.

Becnel, Lezin A. *Report on the Results of Belle Alliance, Evan Hall and Souvenir Sugar Houses, for the Crop of 1888.* New Orleans, 1889.

Boussingault, Jean Baptiste. *Rural Economy, in its relations with chemistry, physics and meteorology.* Translated by George Law. New York, 1845.

Brent, Joseph. *The Lugo Case and Capture of the Indianola.* New Orleans, 1926.

———. *Memoirs of the War Between the States.* New Orleans, 1940.

Browne, C. A., and R. E. Blouin. *The Chemistry of the Sugar Cane and Its*

Products in Louisiana. Louisiana Experiment Station Bulletin XCI. Baton Rouge, 1906.

Burden of the Sugar Tax Protest Against Recommendations of the Tariff Commission. Boston, 1882.

Charter, Constitution, By-Laws and Rules of the Louisiana Sugar and Rice Exchange of New Orleans, Louisiana: Incorporated 6 March 1883. New Orleans [1883].

Charter and Rules of the Boston Club of New Orleans. New Orleans, 1892.

The Chemistry of Sugar Cane and Its Products. Louisiana Experiment Station Bulletin XXXVIII. Baton Rouge, 1895.

Clarke, Isaac Edwards. *Art and Industry: Education in the Industrial and Fine Arts in The United States: Industrial and Technical Training in Schools of Technology and in U.S. Land Grant Colleges.* Washington, 1898.

Clerget, T. *Analyses Des Sucres et Des Substances Saccharifieres.* Paris, 1846.

Collier, Peter. *Report of Analytical and other Work Done on Sorghum and Cornstalks.* Washington, D.C., 1881.

———. *Sorghum as a Source of Sugar: Including a Review of the Bulletins of the Department of Agriculture.* N.p., n.d.

———. *Sorghum, Its Culture and Manufacture Economically Considered as a Source of Sugar, Syrup and Fodder.* Cincinnati, 1884.

———. *Sorghum, Its Culture and Uses: An Address Delivered by Dr. Peter Collier Before the Chamber of Commerce of the State of New York, March 5, 1885.* New York, 1885.

Crampton, Charles Albert. *Record of Experiments at Des Lignes Sugar Experiment Station, Baldwin, La., during the Season of 1888.* U.S.D.A. Division of Chemistry Bulletin XXII. Washington, D.C., 1889.

Cross, W. E. *Clarification of Louisiana Cane Juices.* Louisiana Experiment Station Bulletin CXLIV. Baton Rouge, 1914.

———. *(Technical) Investigations on Methods of Analysis of Cane Products.* Louisiana Experiment Station Bulletin CXXXV. Baton Rouge, 1912.

Davis, George E. *A Handbook of Chemical Engineering.* 2nd ed. 2 vols. Manchester, Eng., 1904.

DeBow, J. D. B. *The Industrial Resources of the Southern and Western States.* 3 vols. New Orleans, 1853.

Deerr, Noel. *Cane Sugar.* London, 1911.

Degrand, E. *Fabrication et Raffinage du sucre: Notice sur la concentration des jus sucres et la cuisson des sirops.* Paris, 1835.

Dumas, Jean Baptiste. *Traite de Chimie Appliquee Aux Arts.* 8 vols. Paris, 1828–46.

Dutrone la Couture, Jacques Francois. *Precis sur la canne et sur les moyens d'en extraire le sel essentiel.* Paris, 1790.

Edson, Hubert. *Calumet Sugarhouse Results, Campaign of 1889–1890.* Louisville, Ky., 1890.

———. *Calumet Plantation Factory Report: Campaign of 1890–1891.* Louisville, Ky., 1891.

———. *Calumet Plantation Report: Campaign 1891–1892.* Louisville, Ky., 1892.

———. *Calumet Plantation Factory Report: Campaign of 1897–1898.* Louisville, Ky., 1898.

———. *Record of Experiments at the Sugar Experiment Station on Calumet Plantation, Pattersonville, La.* U.S.D.A. Division of Chemistry Bulletin XXIII. Washington, 1889.

Edwards and Haubtman. *Illustrated Catalog of Edwards and Haubtman.* New Orleans, 1889.

Evans, William J. *The Sugar-Planters Manual: Being a Treatise on the Art of Obtaining Sugar from the Sugar Cane.* London, 1847.

Extracts of a Few Letters and Notices Received at the Department of Agriculture During the Administration of Gen. Wm. G. LeDuc. Washington, D.C., 1881.

Flint, Timothy. *Recollections of the Last Ten Years.* 1826; rpr. New York, 1968.

Forstall, Edward J. *Agricultural Productions of Louisiana.* N.p., 1845.

Frese, O. *Beitrage zur Zuckerfabrikation . . . Ein Hulfsbuck fur Fabrikherren, Direktoren und Siedemeister.* Brunswick, 1865.

Gilman, Samuel H. *Begass Considered as Fuel for Making Sugar.* New Orleans, 1856.

Goessman, Charles A. *Notes on the Manufacture of Sugar in the Island of Cuba.* Syracuse, N.Y., 1865.

The Government and the Sugar Trade. New York [1878].

Ham, Charles H. *Manual Training: The Solution of Social and Industrial Problems.* New York, 1886.

Hardee, William J. *Hardee's New Geographical, Historical and Statistical Official Map of Louisiana.* Chicago, 1895.

Hausbrand, E. *Evaporating, Condensing and Cooling Apparatus.* Translated by A. C. Wright. 2nd Eng. ed. London, 1916.

Horsin-Dèon, Paul. *Traité Théorique et Pratique de la Fabrication du Sucre: Guide du Chemiste Fabricant.* Paris, 1882.

House Documents. 21st Cong. 2nd Session., No. 62.

Ingersoll, Robert G. *A Review of the Sugar Question.* Washington, D.C., n.d.

Johnson, Samuel W. *How Crops Feed: A Treatise on the Atmosphere and* the Soil as Related to the Nutrition of Agricultural Plants, with Illus-trations. New York, 1882.

———. How Crops Grow: A Treatise on the Chemical Composition, Struc-ture and Life of the Plant, For all Students of Agriculture. New York, 1887.

Kellar, Herbert Anthony, ed. Solon Robinson: Pioneer and Agriculturist, Selected Writings. Vol. II of 2 vols. 1936; rpr. New York, 1968.

Kerr, E. W. Begasse Drying. Louisiana Experiment Station Bulletin CXXVII. Baton Rouge, 1911.

———. Experiments with Oil Burning in Boiler Furnaces. Louisiana Experi-ment Station Bulletin CXXXV. Baton Rouge, 1911.

———. An Experimental Study of Bagasse and Bagasse Furnaces. Louisi-ana Experiment Station Bulletin CXVII. Baton Rouge, 1909.

———. An Experimental Study of Heat Transmission. Louisiana Experi-ment Station Bulletin CXXXVIII. Baton Rouge, 1913.

———. Experimental Studies of Vacuum Juice Heaters. Louisiana Experi-ment Station Bulletin CLIX. Baton Rouge, 1916.

———. Performance Tests of Sugar House Heating and Evaporating Appa-ratus. Louisiana Experiment Station Bulletin CXLIX. Baton Rouge, 1914.

———. Preliminary Tests of Sugar House Machinery. Louisiana Experiment Station Bulletin CVII. Baton Rouge, 1908.

King, J. Floyd. The Government Experiments and the Sugar Trade. Washing-ton, D.C., 1887.

Kingzett, C. T. The History, Productions and Processes of the Alkali Trade. London, 1877.

Kratz, Otto. The Robert Diffusion Process Applied to Sugar Cane in Louisi-ana in the Year 1873 and 1874. New Orleans, 1875.

LeDuc, William G. Recollections of a Civil War Quartermaster: The Auto-biography of William G. LeDuc. St. Paul, Minn., 1963.

Leon, J. A. On Sugar Cultivation. London, 1848.

Liebig, Justus. Chemische Briefe. Leipzig, 1878.

Lomas, John. A Manual of the Alkali Trade. London, 1880.

Loring, George B. The Sorghum Sugar Industry: Address . . . Before the Mississippi Valley Cane Growers' Association, Saint Louis, Mo., De-cember 14, 1882. Washington, D.C., 1883.

Louisiana Board of Commissioners to the Louisiana Purchase Exposition. Louisiana at Louisiana Purchase Exposition. New Orleans, 1904.

Louisiana Bureau of Immigration. An Invitation to Immigrants: Louisiana: Its Products, Soil and Climate as Shown by Northern and Western Men Who Now Reside in this State. Baton Rouge, 1894.

Louisiana State Board of Agriculture and Immigration. *Louisiana's Invitation.* N.p. [1911?].

Louisiana State University Board of Supervisors. *Report . . . to the General Assembly of the State of Louisiana for the Sessions 1896–97 and 1897–98.* Baton Rouge, 1898.

Louisiana Sugar Experiment Station. *Bulletin No. 1, Kenner, La., November, 1885.* New Orleans, n.d.

———. *Bulletin No. 2, . . . January 20, 1886.* N.p., 1886.

———. *Bulletin No. 3, . . . April 1, 1886.* N.p., 1886.

Louisiana Sugar Planters' Association. *Regular Monthly Meeting, May 10, 1884.* N.p., 1884.

———. *Regular Monthly Meeting, May 14, 1885.* N.p., 1885.

———. *Regular Monthly Meeting, July 1887.* New Orleans, 1887.

———. *Regular Monthly Meeting, September 1887.* N.p., 1887.

———. *Regular Monthly Meeting, December 1887.* N.p.,1887.

McCulloch, R. S. *On the Present State of Knowledge of the Chemistry and Physiology of the Sugar Cane and of Saccharine Substances.* New Orleans, 1886.

McDonald, John S. *Automatic Hydraulic Pressure Regulator for Sugar Mills.* New Orleans, 1884.

McMurtrie, William. *Report on the Culture of the Sugar Beet and the Manufacture of Sugar Therefrom in France and the United States.* Washington, D.C., 1880.

Mallet, John William. *Chemistry Applied to the Arts: A Lecture Delivered Before the University of Virginia, May 20, 1868.* Lynchburg, Va., 1868.

———. *Cotton: The Chemical, Geological and Meteorological Conditions Involved in its Successful Cultivation.* London, 1862.

Maxwell, Walter, and Harvey W. Wiley. *Experiments with Sugar Beets in 1892.* U.S.D.A. Division of Chemistry Bulletin XXXVI. Washington, D.C., 1893.

Morrison, J. Chronegh. *Louisiana and its Resources; The State of the Future: An Official Guide for Capitalists, Manufacturers, Agriculturists and the Emigrating Masses.* N.p. [1886?].

National Academy of Sciences. *Investigations of the Scientific and Economic Relations of the Sorghum Sugar Industry . . . November, 1882.* Washington, D.C., 1883.

Olmsted, Frederick Law. *A Journey in the Seaboard Slave States, With Remarks on Their Economy.* 1856; rpr. New York, 1968.

Opening Exercises of the Louisiana Sugar Exchange of New Orleans, 3 June 1884. New Orleans, 1884.

Payen, Anselme. *Cours de chimie elementaire et industrielle.* 2 vols. Paris, 1832–33.

Peligot, Eugène. *Recherches sur la nature et les proprietes chimiques des sucres.* Paris, 1838.

Prinsen Geerlings, H. C. *On Cane Sugar and the Process of its Manufacture in Java.* Manchester, Eng., 1898.

Publica de Cuba. Secretaria de Agricultura. *Comercio Y Trabajo, Industria Azucarera de Cuba, 1913–1914.* N.p., n.d.

Reed, William. *The History of Sugar and Sugar Yielding Plants.* London, 1866.

Report of the Committee on Methods of Sugar Analyses of the Louisiana Sugar Chemists' Association. Baton Rouge [1889].

Rolfe, George William. *The Polariscope in the Chemical Laboratory.* New York, 1905.

Silliman, Benjamin. *Manual on the Cultivation of the Sugar Cane and the Fabrication and Refinement of Sugar.* Washington, D.C., 1833.

[Miles, William Porcher (?)] *Some Views on Sugar: By an Old-Fashioned Democrat.* N.p. [1894].

Spencer, Guilford Lawson. *A Hand-Book for Chemists of Beet Sugar Houses and Seed-Culture Farms: Containing Selected Methods of Analysis, Sugar House Control, Reference Tables.* New York, 1897.

———. *Report of Experiments in the Manufacture of Sugar at Magnolia Station, Lawrence, La., Season of 1885–86: Second Report.* U.S.D.A. Division of Chemistry Bulletin XI. Washington, D.C., 1886.

———. *Report of Experiments in the Manufacture of Sugar by Diffusion at Magnolia Station, La., Season of 1888–1889.* U.S.D.A. Division of Chemistry Bulletin XXI. Washington, D.C., 1889.

Stammer, Karl. *Jahres-Bericht uber die Untersuchungen und Fortschritte auf dem Gesamtgebiet der Zucker Fabrikation.* Breslau, 1868.

———. *Jahres-Bericht uber die Untersuchungen und Fortschritte auf dem Gesamtgebiet der Zucker Fabrikation.* Breslau, 1875.

Statuten des Assekuranzvereins Oesterreichicher Zuckerfabrikanten in Prag. Prag, 1862.

Stillman's Patent Begasse Furnace. New Orleans, 1855.

Stubbs, William Carter. *A Hand-Book of Louisiana . . . Giving . . . Geographical and Agricultural Features.* New Orleans, 1895.

———. *Ramie—. . . Uses, History, . . . With a Report of Committee on the Recent Public Trial of Ramie Machines, at Audubon Park, New Orleans, January 1895.* Louisiana Experiment Station Bulletin XXXII. Baton Rouge, 1895.

———. *Sugar Cane.* Boston, 1903.

———. *Sugar Cane: Sugar House and Laboratory Experiments—1886.* Louisiana Sugar Experiment Station Bulletin X. Baton Rouge, 1886.

———. *Sugar Cane: A Treatise on the History, Botany and Agriculture of*

Sugar Cane, and the Chemistry and Manufacture of its Juices into Sugar and Other Products. New Orleans, 1897.

Stubbs, William Carter, and Herbert Myrick. *The American Sugar Industry: A Practical Manual on the Production of Sugar Beets and Sugar Cane, and on the Manufacture of Sugar Therefrom.* New York, 1899.

Tehauntepec Inter-Ocean Railroad, Mexico. N.p., 1882.

Thorp, Frank Hall. *Outlines of Industrial Chemistry.* New York, 1898.

Tollens, Bernard. *Kurzes Handbuch der Kohlenhydrate.* Vol. II of 2 vols. Breslau, 1895.

True, Alfred Charles, and V. A. Clark. *The Agricultural Experiment Stations in the United States.* Washington, D.C., 1900.

U.S. Congress. House of Representatives. *Hearings Held Before the Special Committee on Investigation of the American Sugar Refining Co. and Others.* 3 vols. Washington, D.C., 1911.

U.S. Department of Agriculture. Division of Chemistry. *Record of Experiments Conducted by the Commissioner of Agriculture in the Manufacture of Sugar from Sorghum and Sugar Canes . . . 1887–1888.* Washington, D.C., 1888.

U.S. Department of Agriculture. Office of Experiment Stations. *Organization of the Agricultural Experiment Stations in the United States: February 1889.* Washington, D.C., 1889.

U.S. Patent Office. *Specifications of Patents.* Washington, D.C., 1876.

U.S. Tariff Commission. *Report.* 2 vols. Washington, D.C., 1882.

U.S. Treasury Department. *The United States vs. 712 Bags of Sugar Imported in the Mississippi.* Document 111. N.p. [1878?].

Ville, George. *Chemical Manures: Agricultural Lectures Delivered at the Experimental Farm at Vincennes, in the Year 1867.* Translated by E. L. Howard. Atlanta, 1871.

——. *The School of Chemical Manures: Or Elementary Principles in the Use of Fertilizing Agents.* Philadelphia, 1872.

Wagner, Rudolf. *A Handbook of Chemical Technology.* Translated by William Crookes. New York, 1892.

Walkhoff, L. *Traite Complet De Fabrication et Raffinage Du Sucre De Betterves.* Translated by E. Merijot and J. Gay-Lussac. 2 vols. Paris, 1875.

Wells, David A. *The Sugar Industry of the United States and the Tariff.* New York, 1878.

Wiles, Robert. *Cuban Cane Sugar.* Indianapolis, 1916.

Wiley, Harvey, W. *An Autobiography.* Indianapolis, 1930.

——. *Diffusion: Its Application to Sugar-Cane, and Record of Experiment with Sorghum in 1883.* U.S.D.A. Division of Chemistry Bulletin II. Washington, D.C., 1884.

———. *Experiments with Diffusion and Carbonatation at Ottawa, Kansas: Campaign of 1885.* U.S.D.A. Division of Chemistry Bulletin VI. Washington, D.C., 1885.

———. *Methods and Machinery for the Application of Diffusion to the Extraction of Sugar from Sugar Cane and Sorghum, and for Use of Lime, and Carbonic and Sulphurous Acids in Purifying the Diffusion Juices.* U.S.D.A. Division of Chemistry Bulletin VIII. Washington, D.C., 1886.

———. *Our Sugar Supply: Annual Address of President of the Chemical Society of Washington.* Washington, D.C., 1887.

———. *Record of Experiments at Fort Scott, Kansas, in the Manufacture of Sugar from Sorghum and Sugar-Canes, in 1886.* U.S.D.A. Division of Chemistry Bulletin XIV. Washington, D.C., 1887.

———. *Record of Experiments with Sorghum in 1891.* U.S.D.A. Division of Chemistry Bulletin XXXIV. Washington, D.C., 1892.

———. *Record of Experiments with Sorghum in 1892.* U.S.D.A. Division of Chemistry Bulletin XXVII. Washington, D.C., 1893.

Wiley, Harvey W. *The Sugar Industry of the United States.* U.S.D.A. Division of Chemistry Bulletin V. Washington, D.C., 1885.

Wilkinson, J. B., Jr. *Wilkinson's Report on Diffusion and Mill Work in the Louisiana Sugar Harvest of 1889–90.* New Orleans, 1890.

Woodward, C. M. *The Manual Training School, Comprising a Full Statement of Its Aims, Methods, and Results.* Boston, 1887.

DISSERTATIONS

Beeson, Jasper Luther. "A Study of the Action of Certain Diazo-Compounds on Methyl and Ethyl Alcohols Under Varying Conditions." Ph.D. dissertation, Johns Hopkins University, 1893.

Coates, Charles Edward. "The Action of Aniline and of Toluidine on Ortho-Sulphobenzoic Acid and its Chloride." Ph.D. dissertation, Johns Hopkins University, 1891.

Secondary Sources

BOOKS

Aldrich, Howard E. *Organizations and Environments.* Englewood Cliffs, N.J., 1979.

Anderson, Oscar E. *The Health of a Nation: Harvey Wiley and the Fight for Pure Food.* Chicago, 1958.

Arthur, Stanley C., and George de Kernion. *Old Families of Louisiana.* New Orleans, 1931.

Aykroyd, W. R. *Sweet Malefactor.* London, 1967.

Baker, Gladys L., Wayne D. Rasmussen, Vivian Wiser, and Jane L. Porter. *Century of Service: The First 100 Years of the United States Department of Agriculture.* Washington, D.C., 1963.

Berlin, Ira. *Slaves Without Masters: The Free Negro in the Antebellum South.* New York, 1974.

Billings, Dwight B., Jr. *Planters and the Making of a New South.* Chapel Hill, 1979.

Biographical and Historical Memoirs of Louisiana. 2 vols. Chicago, 1892.

Browne, C. A., and F. W. Zerban. *Physical and Chemical Methods of Sugar Analysis.* New York, 1941.

Butler, Pierce. *Judah P. Benjamin.* Philadelphia, 1906.

Christian, Marcus. *Negro Ironworkers in Louisiana 1718–1900.* Gretna, La., 1972.

Clapham, John. *The Economic Development of France and Germany, 1815–1914.* Cambridge, Eng., 1921.

Clemens, Samuel Langhorne. *Life on the Mississippi.* New York, 1917.

Cobb, James C. *Industrialization and Southern Society, 1877–1984.* Lexington, Ky., 1984.

Colman, Gould P. *Education and Agriculture: A History of the New York State College of Agriculture at Cornell University.* Ithaca, N.Y., 1963.

Cummins, Light Townsend, and Glen Jeansonne, eds. *A Guide to the History of Louisiana.* Westport, Conn., 1982.

Dalton, John E. *Sugar: A Case Study of Government Control.* New York, 1937.

Deerr, Noel. *The History of Sugar.* 2 vols. London, 1949–50.

Desdunes, R. L. *Nos Hommes Et Notre Histoire.* Montreal, 1911.

Dupree, A. Hunter. *Science in the Federal Government: A History of Policies and Activities to 1940.* New York, 1964.

Dyer, John P. *Tulane: The Biography of a University.* New York, 1966.

Edson, Hubert. *Sugar—From Scarcity to Surplus.* New York, 1958.

Eichner, Alfred S. *The Emergence of Oligopoly: Sugar Refining as a Case Study.* Baltimore, 1969.

Eisenberg, Peter L. *The Sugar Industry in Pernambuco: Modernization Without Change, 1840–1910.* Berkeley, 1974.

Ellis, E. *An Introduction to the History of Sugar as a Commodity.* Philadelphia, 1905.

Ellis, Lippert S. *The Tariff on Sugar.* Freeport, Ill., 1933.

Fleming, Walter L. *Louisiana State University, 1860–1896.* Baton Rouge, 1936.

Furter, William F., ed. *A Century of Chemical Engineering.* New York, 1982.

———. *History of Chemical Engineering.* Washington, D.C., 1980.

Gray, Lewis Cecil. *History of Agriculture in the Southern States to 1860.* Vol. II of 2 vols.: 1933; rpr. Clifton, N.J., 1973.

Greathouse, Charles H. *Historical Sketch of the U.S. Department of Agriculture: Its Objects and Present Organization.* Washington, D.C., 1898.

Hair, William Ivey. *Bourbonism and Agrarian Protest: Louisiana Politics, 1877–1900.* Baton Rouge, 1969.

Harbison, David. *Reaching for Freedom: Paul Cuffe, Norbert Rillieux, Ira Aldridge, James McCune Smith.* N.p., 1972.

Hardy, D. Clive. *The World's Industrial and Cotton Centennial Exposition.* New Orleans, 1978.

Hughes, Thomas P. *Networks of Power: Electrification in Western Society.* Baltimore, 1983.

Ihde, Aaron J. *The Development of Modern Chemistry.* New York, 1964.

Laitinen, Herbert A., and Walter E. Harris. *Chemical Analysis.* New York, 1975.

Landry, Stuart O. *History of the Boston Club.* New Orleans, 1938.

Laughlin, Clarence John. *Ghosts Along the Mississippi.* New York, 1961.

Lemmer, George F. *Norman J. Colman and Colman's Rural-World.* Columbia, Mo., 1953.

Lippmann, Edmund O. von. *Geschichte des Zuckers.* 2 vols. Berlin, 1929.

Lohn, Ella. *Reconstruction in Louisiana after 1868.* New York, 1930.

Massachusetts Agricultural College. *Charles Anthony Goessman.* Cambridge, Mass., 1917.

Meade, Robert Donthat. *Judah P. Benjamin.* New York, 1943.

Mintz, Sidney W. *Sweetness and Power: The Place of Sugar in Modern History.* New York, 1985.

Moody, V. Alton. *Slavery on Louisiana Sugar Plantations.* N.p. [1924?].

Nau, John Frederick. *The German People of New Orleans, 1850–1900.* Leiden, 1958.

Neiman, Simon I. *Judah Benjamin.* Indianapolis, 1963.

Officers, Members, Charter and Rules of the Boston Club. N.p., 1966.

Prescott, Samuel C. *When M.I.T. was "Boston Tech."* Cambridge, Mass., 1954.

Reed, Germaine M. *David French Boyd.* Baton Rouge, 1977.

Reynolds, Terry S. *75 Years of Progress: A History of the American Institute of Chemical Engineers, 1908–1983.* New York, 1983.

Rosenberg, Nathan. *Technology and Economic Growth.* New York, 1972.

Rossiter, Margaret. *The Emergence of Agricultural Science: Justus Liebig and the Americans.* New Haven, Conn., 1975.

Rousseve, Charles Barthelmy. *The Negro in Louisiana.* New Orleans, 1937.

Scarano, Francisco A. *Sugar and Slavery in Puerto Rico: The Plantation Economy of Ponce, 1800–1850.* Madison, Wis., 1984.

Schmitz, Mark. *Economic Analysis of Antebellum Sugar Plantations in Louisiana*. New York, 1977.

Schuchart, Theodor. *Die Volkswirtschaftliche Bedeutung der Technischen Entwicklung der Deutschen Zukerindustrie*. Leipzig, 1908.

Sitterson, J. Carlyle. *Sugar Country: The Cane Sugar Industry in the South, 1753–1950*. Lexington, Ky., 1953.

Skipper, Ottis Clark. *J. D. B. DeBow, Magazinist of the Old South*. Athens, Ga., 1958.

Starobin, Robert S. *Industrial Slavery in the Old South*. New York, 1970.

Sterkx, H. E. *The Free Negro in Ante-Bellum Louisiana*. Rutherford, N.J., 1972.

Taylor, Joe Gray. *Louisiana Reconstructed*. Baton Rouge, 1974.

——. *Negro Slavery in Louisiana*. New York, 1969.

True, Alfred Charles. *A History of Agricultural Experimentation and Research in the United States, 1607–1925*. U.S.D.A. Miscellaneous Publication CCLI. Washington, D.C., 1937.

——. *A History of Agricultural Education in the United States, 1785–1925*. Washington, D.C., 1929.

Van Antwerpen, F. J., and Sylvia Fourdrinier. *Highlights of the First Fifty Years of the American Institute of Chemical Engineers*. N.p., 1958.

Warmoth, Henry Clay. *War, Politics and Reconstruction*. New York, 1930.

Warren, Kenneth. *Chemical Foundations: The Alkali Industry in Britain to 1926*. Oxford, 1980.

Weber, Gustavus A. *The Bureau of Chemistry and Soils: Its History, Activities and Organization*. Baltimore, 1928.

Whitten, David O. *The Emergence of Giant Enterprise, 1860–1914*. Westport, Conn., 1983.

Wiebe, Robert. *The Search for Order, 1877–1920*. New York, 1967.

Wiener, Jonathan M. *Social Origins of the New South: Alabama, 1860–1885*. Baton Rouge, 1978.

Wik, Reynold M. *Steam Power on the American Farm*. Philadelphia, 1953.

Wilkerson, Marcus M. *Thomas Duckett Boyd*. Baton Rouge, 1935.

Woodward, C. Vann. *Origins of the New South, 1877–1913*. Baton Rouge, 1951.

ARTICLES AND ESSAYS

Amundson, Richard J. "Oakley Plantation: A Post–Civil War Venture in Louisiana Sugar." *Louisiana History*, IX (1968), 21–42.

Arsenault, Raymond. "The End of the Long Hot Summer: The Air Conditioner and Southern Culture." *Journal of Southern History*, L (1984), 597–628.

Browne, Charles A. "Bernard Tollens (1841–1918) and Some American Students of His School of Agricultural Chemistry." *Journal of Chemical Education*, XIX (1942), 253–59.

———. "Origin of the Clerget Method." *Chemical and Engineering News*, XX (1942), 322–24.

Destler, Chester McArthur. "David Dickson's 'System of Farming' and the Agricultural Revolution in the Deep South, 1850–1885." *Agricultural History*, XXXI (1957), 30–39.

Ferleger, Louis. "Farm Mechanization in the Southern Sugar Sector After the Civil War." *Louisiana History*, XXIII (1982), 21–34.

Fox, William Lloyd. "Harvey W. Wiley's Search for American Sugar Self-Sufficiency." *Agricultural History*, LIV (1980), 516–26.

Galambos, Louis. "Technology, Political Economy, and Professionalization: Central Themes of the Organizational Synthesis." *Business History Review*, LVII (1983), 471–93.

Gayarré, Charles. "A Louisiana Sugar Plantation of the Old Regime." *Harper's New Monthly Magazine*, LXXIV (1887), 606–21.

Hackney, Sheldon. "Origins of the New South in Retrospect." *Journal of Southern History*, XXXVIII (1972), 191–216.

Harris, Francis Byers. "Henry Clay Warmoth, Reconstruction Governor of Louisiana." *Louisiana Historical Quarterly*, XXX (1947), 523–63.

Le Gardeur, Rene J., Jr. "The Origins of the Sugar Industry in Louisiana." In Center for Louisiana Studies, *Green Fields: Two Hundred Years of Louisiana Sugar*, Lafayette, La., 1980.

Lewis, W. K. "Evaluation of Unit Operations." *Engineering Progress*, LV (1959), Symposium Series No. 26.

Nash, Gerald D. "Research Opportunities in the History of the South After 1880." *Journal of Southern History*, XXXII (1966), 308–24.

Oesper, Ralph E. "Kjeldahl and the Determination of Nitrogen." *Journal of Chemical Education*, XI (1934), 457–62.

Porter, Jane M. "Experiment Stations in the South, 1877–1940." *Agricultural History*, LIII (1979), 84–100.

Reed, Merl E. "Footnote to the Coastwise Trade—Some Teche Planters and Their Atlantic Factors." *Louisiana History*, VIII (1967), 191–97.

Reich, Leonard S. "Industrial Research and the Pursuit of Corporate Security: The Early Years of Bell Labs." *Business History Review*, LIV (1980), 504–29.

———. "Irving Langmuir and the Pursuit of Science and Technology in the Corporate Environment." *Technology and Culture*, XXIV (1983), 199–221.

———. "Research, Patents and the Struggle to Control Radio: A Study of Big Business and the Uses of Research." *Business History Review*, LI (1977), 208–35.

Rogers, Benjamin F., Jr. "The United States Department of Agriculture and the South, 1862–1880." *Studies* (Florida State University), X (1953), 71–80.

Roland, Charles P. "Difficulties of Civil War Sugar Planting in Louisiana." *Louisiana Historical Quarterly*, XXXVIII (1955), 40–62.

Rosenberg, Charles E. "The Adams Act: Politics and the Cause of Scientific Research." *Agricultural History*, XXXVIII (1964), 3–21.

———. "Science, Technology and Economic Growth: The Case of the Agricultural Experiment Station Scientist, 1875–1914." *Agricultural History*, XLIV (1971), 1–20.

Saloutos, Theodore. "The Grange in the South, 1870–77." *Journal of Southern History*, XIX (1953), 473–87.

Scarpaci, J. Vincenza. "Labor from Louisiana's Sugar Cane Fields: An Experiment in Immigrant Recruitment." *Italian Americana*, VII (1981), 19–41.

Servos, John W. "The Industrial Relations of Science: Chemical Engineering at MIT, 1900–1939." *Isis*, LXXI (1980), 531–49.

Sitterson, J. Carlyle. "Business Leaders in Post–Civil War North Carolina, 1865–1900." In *Studies in Southern History*, edited by Sitterson, Chapel Hill, 1957.

———. "Expansion, Reversion, and Revolution in the Southern Sugar Industry, 1850–1910." *Bulletin of the Business History Society*, XXVII (1953), 129–40.

———. "Financing and Marketing the Sugar Crop of the Old South." *Journal of Southern History*, X (1944), 188–99.

———. "Magnolia Plantation, 1852–1862." *Mississippi Valley Historical Review*, XXV (1938–39), 197–210.

Thomas, Milton Halsey. "Professor McCulloh of Princeton, Columbia and Points South." *Princeton University Library Chronicle*, IV (1947), 17–29.

Toledano, Roulhac B. "Louisiana's Golden Age: Valcour Aime in St. James Parish." *Louisiana History*, X (1969), 211–24.

White, Alfred H. "Chemical Engineering Education in the United States." *Transactions of the American Institute of Chemical Engineers*, XXI (1928), 55–85.

Wiener, Jonathan M. "Class Structure and Economic Development in the American South, 1865–1955." *American Historical Review*, LXXXIV (1979), 970–93.

DISSERTATIONS AND OTHER UNPUBLISHED MATERIALS

Binning, Francis Wayne. "Henry Clay Warmoth and Louisiana Reconstruction." Ph.D. dissertation, University of North Carolina at Chapel Hill, 1969.

Foote, Irving P. "A Louisiana Sugar Plantation and the Sugar Industry: A Type Study." M.A. thesis, George Peabody College for Teachers [1922(?)].

Heitmann, John. "The British Alkali Industry, 1860–1890, and the Emergence of Chemical Engineering." Paper, Johns Hopkins University, 1979.

———. "Two Examples of Early Chemical Engineering Education in America: The Audubon Sugar School and the University of Wisconsin." Paper delivered at Society of History of Technology Annual Meeting, 1980.

Index

291

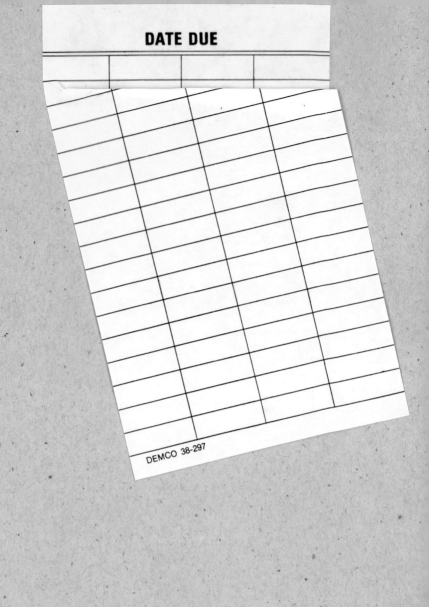

DATE DUE

DEMCO 38-297